THE PROMISE OF INFRASTRUCTURE

THE PROMISE OF INFRASTRUCTURE

Nikhil Anand, Akhil Gupta, and Hannah Appel, editors

A School for Advanced Research Advanced Seminar

———

DUKE UNIVERSITY PRESS   DURHAM AND LONDON   2018

Designed by Courtney Leigh Baker
Typeset in Garamond Premier Pro and Univers by
Westchester Publishing Services

Library of Congress Cataloging-in-Publication Data
Names: Anand, Nikhil, [date] editor. | Gupta, Akhil, [date] editor. |
Appel, Hannah, [date] editor.
Title: The promise of infrastructure / edited by Nikhil Anand,
Akhil Gupta, and Hannah Appel.
Description: Durham : Duke University Press, 2018. | "A School for
Advanced Research Advanced Seminar." | Includes bibliographical
references and index.
Identifiers: LCCN 2017058620 (print)
LCCN 2018000308 (ebook)
ISBN 9781478002031 (ebook)
ISBN 9781478000037 (hardcover : alk. paper)
ISBN 9781478000181 (pbk. : alk. paper)
Subjects: LCSH: Infrastructure (Economics)—Social aspects. |
Infrastructure (Economics)—Political aspects. | Economic
development—Social aspects. | Technological complexity—
Social aspects. | Ethnology. | Technology—Social aspects.
Classification: LCC HC79.c3 (ebook) | LCC HC79.c3 P78 2018 (print) |
DDC 363.6—dc23
LC record available at https://lccn.loc.gov/2017058620

Cover art: *Bridge Pile Reinforcement Structure*. Zhengzaishuru/
Shutterstock.com.

CONTENTS

ACKNOWLEDGMENTS

This book began as an Advanced Seminar at the School for Advanced Research (SAR) in Santa Fe, New Mexico. We thank Nicole Taylor, Leslie Shipman, and Michael Brown for tremendous work in creating and maintaining its special scholarly environment. We were encouraged and inspired by the wonderful spirit of collaboration of our seminar participants: Geoffrey Bowker, Dominic Boyer, Catherine Fennell, Penny Harvey, Brian Larkin, Antina von Schnitzler, and Christina Schwenkel. Unfortunately, due to unforeseen circumstances, Catherine Fennell needed to withdraw her contribution from the volume. Nevertheless her insights continue to animate the book. We are also indebted to the participants on two double panels focused on the anthropology of infrastructure at the annual meetings of the American Anthropological Association, first in New Orleans in November, 2010, and subsequently in Chicago in November, 2013. We would like to thank Jessica Barnes, Andrea Ballestero, Dominic Boyer, Ashley Carse, Brenda Chalfin, Penny Harvey, Brian Larkin, Michael Degani, Curt Gambetta, Elizabeth Povinelli, Janell Rothenberg, Julia Elyachar, Bill Maurer, Suzana Sawyer, Antina von Schnitzler, K. Sivaramakrishnan, and Gisa Weszkalnys for their generous contributions at these sessions. The opportunity to think with Arjun Appadurai, Jessica Barnes, Laura Bear, Ashley Carse, and Austin Zeiderman in these and other fora was especially key in conceptualizing this project. We are grateful for the feedback we received when we presented this work at the Studio for Ethnographic Design workshop at the University of California at San Diego and at the Institute for Advanced Study at the University of Minnesota. Finally, our thanks to Ken Wissoker, Elizabeth Ault, and the editorial team at Duke for their enthusiasm for this project and for their hard work in bringing it out to the world.

Temporality, Politics, and the Promise of Infrastructure

HANNAH APPEL, NIKHIL ANAND,
AND AKHIL GUPTA

The settlers' town is a strongly built town, all made of stone and steel. It is a brightly lit town; the streets are covered with asphalt, and the garbage cans swallow all the leavings, unseen, unknown and hardly thought about. . . . The town belonging to the colonized people . . . is a world without spaciousness; men live there on top of each other, and their huts are built one on top of the other. The native town is a hungry town, starved of bread, of meat, of shoes, of coal, of light.

—FRANTZ FANON (1961)

In April 2014, the Detroit Water and Sewage Department began to turn off the water of city residents who were behind on their payments. Over the course of the year, tens of thousands of Detroit inhabitants lost access to water in their homes, and fought furiously for its restoration. Also in April 2014, seventy miles north, the city of Flint, Michigan, switched its water source from Lake Huron to the Flint River. Polluted with the effluent from heavy industry and toxic bacteria, chloride and chlorine-based disinfectants that were intended to make the water drinkable only exacerbated the problem. Without additional chemicals to ensure that the first set of chemical additives would not disintegrate pipes, the treated river water corroded the aging plumbing infrastructure made of copper, iron, and lead. Heavy metals leached into municipal drinking water, resulting in widespread lead poisoning that was concentrated in children, as well as an outbreak of Legionnaires' disease.

Both the water cutoffs in Detroit and the new water source in Flint occurred at the order of a state-appointed emergency manager, tasked explicitly with austerity and given the authority to override elected officials. The role of the emergency manager is codified in law under Michigan's "Local Financial Accountability and Choice Act" (Act 436 of 2012: §9(2)). After a section ensuring the ongoing provision of services essential to public health, safety, and welfare, the Act reads, "The financial and operating plan shall provide for all of the following: . . . The payment in full of the scheduled debt service requirements on all bonds, notes, and municipal securities of the local government, contract obligations in anticipation of which bonds, notes, and municipal securities are issued, and all other uncontested legal obligations" (§11(1)(b)). Debt servicing on infrastructure, in other words, openly superseded democratic governance. Residents of Flint brought bottles of brown water to city hall meetings and documented rashes and hair loss in children, only to be told repeatedly that the water was safe. The local General Motors plant had already stopped using city water in October 2014, citing concerns about corrosion. When the city council finally voted in March 2015 to reconnect to a safe water source, the emergency management team overruled the vote.

Detroit and Flint are predominantly black cities. Here, water infrastructure is a sociomaterial terrain for the reproduction of racism, which Ruth Wilson Gilmore defines as "the state-sanctioned or extralegal production and exploitation of group-differentiated vulnerability to premature death" (2007: 28). The racial necropolitics of Michigan show that infrastructure is a terrain of power and contestation: To whom will resources be distributed and from whom will they be withdrawn? What will be public goods and what will be private commodities, and for whom? Which communities will be provisioned with resources for social and physical reproduction and which will not? Which communities will have to fight for the infrastructures necessary for physical and social reproduction? In Detroit and Flint, the centrality of municipal debt, the privatization of public goods and services, and market-led governance might lead us to view both as typical examples of neoliberalism in practice. But in both cities, the water shutoffs were as much a result of longer struggles that date back to the 1930s as they were of more immediate pressures emanating from Wall Street (Cramer and Katsarova 2015). For example, one can trace differences in transport and utility infrastructures to the Federal Housing Authority's redlined maps that ensured racial segregation occurred within city limits (Highsmith 2015; Ranganathan 2016). The state-sanctioned infrastructural abandonment that ensued over the following decades is coded today as the product of financially irresponsible residents on whom austerity and dispossession can justly be visited. Indeed, part of Detroit's bankruptcy agreement was the quiet transfer of control of the municipal water infrastructure—serving a largely African American population and governed by their elected leaders—to a regional water board serving Detroit's suburbs. After five decades of deindustrialization and outmigration from the city, the water-board members were still 80 percent white.

Infrastructures around the world—from the United States to Fanon's Algeria to Palestine—offer archaeologies of differential provisioning that predate neoliberalism. Palestine, as Stephen Graham (2002), Eyal Weizman (2012), and others have shown, is a zone of infrastructural warfare, where water tanks, electricity transformers, roads, electronic communications, radio transmitters, and airport runways are often targets. And as the quote from Fanon with which we open this introduction suggests, the experience of infrastructure has long been an affective and embodied distinction between the settlers' town and the town belonging to the colonized people (see also Mrázek 2002; Barak 2013).

We start a volume on the promise of infrastructure here—in the United States, in Palestine, in colonial Algeria—to show the multivalent political trajectories of both infrastructure and the idea of promise. Material infrastructures, including roads and water pipes, electricity lines and ports, oil pipelines and sewage systems, are dense social, material, aesthetic, and political formations that are critical both to differentiated experiences of everyday life and to expectations of the future. They have long promised modernity, development, progress, and freedom to people all over the world. As deep-water rigs drill for oil in West Africa, as roads in Peru or Bangalore promise new connections, or as emerging economies rapidly build dams to modernize their agriculture, infrastructures are critical locations through which sociality, governance and politics, accumulation and dispossession, and institutions and aspirations are formed, reformed, and performed. At the same time as they promise circulation and distribution, however, these precarious assemblies also threaten to break down and fail. From the Deepwater Horizon conflagration to the Fukushima Daiichi nuclear-power plant, from the collapse of school buildings in China to the destruction in the wake of Hurricanes Katrina and Sandy, to the failure of the derivatives market in the 2008 financial crisis, infrastructural breakdown saturates a particular politics of the present. On the one hand, governments and corporations point to infrastructural investment as a source of jobs, market access, capital accumulation, and public provision and safety. On the other hand, communities worldwide face ongoing problems of service delivery, ruination, and abandonment, and they use infrastructure as a site both to make and contest political claims. As the black cities of Michigan or the rubble in Palestine forcefully show, the material and political lives of infrastructure frequently undermine narratives of technological progress, liberal equality, and economic growth, revealing fragile and often violent relations between people, things, and the institutions that govern or provision them. This tension—between aspiration and failure, provision and abjection, and technical progress and its underbelly—makes infrastructure a productive location to examine the constitution, maintenance, and reproduction of political and economic life. What do infrastructures promise? What do infrastructures do? And what does attention to their lives—their construction, use, maintenance, and breakdown; their poetics, aesthetics, and form—reveal?

In recent years, cultural anthropologists have asked these questions of infrastructure. As a result, infrastructure is no longer invoked only as a conceptual tool, as, for instance, in Louis Althusser's (1969) famous invocation of infrastructure in theorizing capitalism, but as itself the object of ethnographic engagement (see Larkin 2013 for a helpful overview). This volume shows how oil rigs and electrical wires, roads and water pipes, bridges and payment systems articulate social relations to make a variety of social, institutional, and material things (im)possible. These "hard" infrastructures are classically anthropological subjects, because attention to them is also attention to sociality, to the ways infrastructure "attracts people, draws them in, coalesces and expends their capacities. . . . People work on things to work on each other, as these things work on them" (Simone 2012).

Infrastructure, like the state in an earlier theoretical moment, has often lurked in the background of anthropological research. Why the surge in contemporary interest? As Antina von Schnitzler asks, "Why infrastructure, why now, and to what end" (von Schnitzler 2015; see also Boyer, this volume)? To answer this question, von Schnitzler attends to the ways in which apartheid was enacted in South Africa through the differential management of infrastructure. Ethnographic attention to infrastructure reveals how politics not only is formed and constrained by juridico-political practices, but also takes shape in a technopolitical terrain consisting of pipes, energy grids, and toilets. An attention to infrastructure, von Schnitzler argues, is classically anthropological because it provides a frame to defamiliarize and rethink the political. Yet to recognize why infrastructure has emerged anew as an analytic and ethnographic object at this moment, we also need to attend to infrastructure's performance as a technology of liberal rule.

In his book *The Rule of Freedom* (2003), the historian Patrick Joyce demonstrates how the construction and management of infrastructure emerged as a key technology of government that was central to the performance of liberalism (see also Mitchell 2011). While privileging the circulation of people and things, infrastructures also served to permit states to separate politics from nature, the technical from the political, and the human from the nonhuman. Thus depoliticized, the management of infrastructures as a technical problem formed the grounds on which subjects were "freed" to participate in civil society and produce economic life. Infrastructures gave form to relations between states and subjects on one hand, and corporations and capitalist circuits on the other. Infrastructures have continued to be central to the work of government since the nineteenth century, and as these opening references to the infrastructures of colonial Algeria, the contemporary United States, and Palestine demonstrate, this form of governance known as liberalism must always be understood, from its inception, as guaranteeing the liberties of some through the subordination, colonization, and racialization of others (Singh 2005; Melamed 2006; Sheth 2009; Mills 2011). Infrastructures have been technologies that modern states use not only to demonstrate development, progress, and modernity, giving these categories their aesthetics, form, and substance (Larkin, this volume), but also to differentiate populations and subject some to premature death (Fanon 1961; Gilmore 2007; McKittrick 2011).

As liberal modernity has (partly) shifted to neoliberal postmodernity, proponents of neoliberalism have argued that particular kinds of infrastructures are necessary to capitalism, and, as such, need to be continually produced (by states, corporations, or different combinations of these) to ensure the reliability of capital and labor flows. Today, as nation-states, particularly in the global South, seek to change their terms of integration into the global economy, they have undertaken dramatic infrastructure projects in varied financial and engineering relationships with private firms. Largely due to foreign investment in its infrastructure, the small central African country of Equatorial Guinea had the highest ratio of investment to gross domestic product of any national economy in the world in 2013 (Harrison 2013). In India, China, and elsewhere, governments see the construction of roads and telecommunication systems as being essential for the production of goods and services for markets as distant as North America and sub-Saharan Africa. The rapid construction of infrastructures in these nation-states is, in turn, shifting the geography of infrastructure expertise. Chinese and Indian companies can be found throughout continental Africa and South America, exporting labor, capital, and inputs like steel to build infrastructures far beyond their national borders.

The uneven flurry of infrastructural investment in the global South coexists with its mirror image in the United States and the United Kingdom, where neoliberal austerity regimes have withdrawn public funds for building and maintaining infrastructure. Such regimes subsist by wearing down the Keynesian investment in the roads, railways, water lines, sewage systems, and telecommunication systems of an earlier historical moment. In the absence of maintenance work on one hand, and neoliberal refigurations of infrastructure grids on the other, existing infrastructures have deteriorated to such an extent that they are breaking down more often (Bennett 2010). This moment has made infrastructure visible in the global North in different ways. As Dominic Boyer argues in this volume,

> The Keynesianism that preceded neoliberalism, dominating western political economic theory and policy from roughly the mid-1930s until the mid-1970s, often utilized large-scale public works projects as key instruments for managing labor, "aggregate demand," and the affective ties of citizenship. Thirty years of privatization, financialization, and globalization later, this legacy of "public infrastructure" has become rather threadbare, capturing a general sense of evaporating futurity in the medium of corroded pipes and broken concrete. Of course, neoliberalism did promote aggressive investment and innovation in infrastructural systems necessary for the advance of financialization and globalization (not least telecommunications, the Internet, and transportation). At the same time, infrastructural temporalities look rather different from the perspective of the global South where . . . ruination is a constant companion of infrastructure. But across the global North, one cannot be faulted for feeling a creeping sense of decay spreading across many infrastructural environments. Thus, the turn to infrastructure could be viewed as something like a conceptual New Deal for the human sciences—a return of the repressed concerns of public developmentalism to an academic environment that has, like much of the rest of the world, become saturated with market-centered messages and logics over the past three decades.

Boyer encourages us to look beyond this nostalgia for petro-fueled Keynesianism, to see the infrastructural turn instead as part of the wider anti-anthropocentric turn in the human sciences. He takes the turn to infrastructure as a sign "that we are conceptually re-arming ourselves for the struggle against the Anthropocene and the modernity that made it," a provocation that we return to later in this introduction.

Within and beyond the histories of (neo)liberalism we describe, infrastructure is an integral and intimate part of daily social life: it affects where and how we go to the bathroom; when we have access to electricity or the Internet; where we can travel, how long it takes, and how much it costs to get there; and how our production and consumption are provisioned with fuel, raw materials, and transport. It is important to underline what may seem self-evident: infrastructures shape the rhythms and striations of social life. Class, gender, race, and kinship are all refracted through differentiated access to infrastructure, deciding whether water or electricity is available and to whom (Ferguson 2012). Who, in a given family or community, carries water from the stream or from a communal tap into the home? Which families can afford a rooftop diesel generator? Cellular networks also reshape gendered socialities: daughters-in-law in Delhi may be allowed to leave the home but their movements are monitored by calls every few minutes; FaceTime and WhatsApp change forms of familial connection and communication at a distance. But even these insights provoke more questions than answers. Take electricity, for example: apart from the fact that people make illegal connections, we know very little of how electricity is actually used within homes. For what do people use electricity? What uses do they consider essential (Degani 2013; Kale 2014)? How is electricity integrated into people's daily lives, from homework for the kids to entertainment and leisure activities (TV, radio, computer, Internet, etc.)? Even as utilities and governments perceive a growing need to handle shortages and to imagine energy transitions, they know very little about daily use, daily need, and what might be socially possible. The promise of infrastructure, then, is multivalent. This volume indexes not only radical disconnection and abandonment, but also aspiration, the prospective, and futurity, of both infrastructures themselves and our work with them. We present a set of scholars working on infrastructure today, but we also gesture to all the work still to be done.

Of course, any given future is built on a past. The relationship between infrastructure, environment, and modernity has preoccupied anthropology since the beginning of the discipline. Cultural materialists like Leslie White (1943), Marvin Harris (1966), Julian Steward (1955), and even Marcel Mauss (2008) were critical of modernization stories of lag and lack, often told through gestures to the technological sophistication of what were believed to be discrete cultures. These theorists paid close attention to the ways in which irrigation, energy, and other technical systems mediated relations among local environments and labor and cultural practices. In his famous consideration of wet and dry irrigation technologies, for instance, Clifford Geertz (1972) drew attention to the ways in which these produced different kinds of persons and political authorities. Engaged with the work of Geertz, Stephen Lansing (1991) attended more closely to the "engineered landscapes" of irrigation in Bali, demonstrating how these infrastructure regimes transform and humanize nature, generating durable political institutions.

Newer work has further developed this engagement by attending to the ways in which environments and landscapes have to be remade so that infrastructures may behave according to human designs. Such projects to manage and order landscapes are always provisional achievements, dependent on the reliable performances of people and environments that are not always under the control of engineers and planners (Ballestero 2015). For instance, Ashley

Carse (2014) demonstrates how distant watersheds in upland Panamanian forests need to be continually made and extended as, in effect, the infrastructure of infrastructure, so that these deliver reliable quantities of water for the Panama Canal. As the materiality of the earth, the reliability of rain, and the political claims on the watershed are variable, the efficacy of the canal depends on the degree to which engineers, hydrologists, and politicians can consistently mobilize the water that it needs to work. Yet, to what degree might we expect nature to continue serving as infrastructure's infrastructure (Jensen 2017)? As humans intervene in the climatic, geological, and evolutionary processes of the Anthropocene (Chakrabarty 2009), both the effects and futures of modern infrastructuring projects appear increasingly tenuous.

This volume is indebted to earlier approaches in the field, not just in anthropology, but also in urban geography and STS. Thus, before turning to the three interventions of the volume (on time, politics, and promise), we schematically lay out an array of genealogies on which this volume builds: (1) critical Marxist perspectives from Althusser to Walter Benjamin, and the development studies literature they have influenced; (2) the government of difference in cities; and (3) the STS literature that attends to the practice of design and engineering. Part of the intervention of this volume, and the emergent anthropology of infrastructure of which it is a part, is to ask how these genealogies can be repurposed to new ethnographic, political, and theoretical ends.

Marxism, Development, and the Telos of Infrastructure

Seeking to account for the social and political changes brought about by the industrial revolution, Marxist and liberal theorists alike often deployed metaphors of infrastructure and technology to make their cases. Take, for instance, Marxist references to infrastructure in theorizing capitalism. In a famous passage, Althusser writes, "Marx conceived the structure of every society as constituted by 'levels' or 'instances' articulated by a specific determination: the infrastructure, or economic base (the 'unity' of the productive forces and the relations of production) and the superstructure, which itself contains two 'levels' or 'instances': the politico-legal (law and the State) and ideology (the different ideologies, religious, ethical, legal, political, etc.)" (1961: 134,). Althusser specifies that his invocation of infrastructure is a metaphor: "Like every metaphor, this metaphor suggests something, makes something visible. What? Precisely this: that the upper floors could not 'stay up' (in the air) alone, if they did not rest precisely on their base. Thus the object of the metaphor of the edifice is to represent above all the 'determination in the last instance' by the economic base" (135). Althusser's famous metaphor of the edifice draws on the meaning of the prefix "infra" (below, beneath, or within) to make an argument about relative autonomy, reciprocal action, and determination in the last instance between infra- and superstructure.

Searching early writing in social thought for more literal accounts of infrastructure as a material form as opposed to a heuristic device, we note that infrastructure often appears as a temporal marker in the techno-developmentalist teleologies (Engels [1884] 2010) that not only animated Marxist approaches to capitalism and theories of economic modernization (Rostow 1960), but also played a similar role in early anthropological theory. For instance, in his attempt to place different cultures in a larger common humanity, Lewis Henry Morgan ([1877] 2004) saw technological development as the force behind cultural development, suggesting that changes in social institutions, organizations, and ideologies emanated from advances in technology. A culture's "arrival" at each progressive stage was marked by a signature technological achievement: fire, bow and arrow, irrigated agriculture, iron manufacture, and so on. Infrastructures and technologies here are both material and symbolic, standing in for a culture (or an economy's) development along a linear temporal scale. Within these now-dismissed theories of teleological progression, we can find the seeds of analytic insight. Take Benjamin's "Iron Construction F" from the Arcades Project. Even as he partakes in developmentalist ideas about the stage of civilization marked by iron, Benjamin also draws our attention to the indivisibility of the "politics and poetics" of infrastructure (Larkin 2013) and to the ways that materials are always "in the grip of dreams" (152) and come with "the peculiar and unmistakable dream world that attaches to them" (156). The tensile properties of iron permit it to be drawn into fantastic material formations: high-rises, arcades, and bridges; formations that celebrate a release from the earth and its histories, gesturing instead to a time and space oriented to the future.

In the late twentieth century, materialist approaches to development withered under poststructural critique, particularly in anthropology (Marcus and Fisher 1986; Gupta and Ferguson 1992). Anthropologists drew on the work of Michel Foucault to argue that material reality does not exist independently of or prior to representational practices. Discourses, narratives, and language give form to infrastructure as much as concrete, wires, or zoning regulations (Ferguson 1994; Escobar 1995). Anthropologists also drew attention to the multiple histories, geographies, and temporalities in relation to which states, infrastructures, and their developmentalist projects were situated (Gupta 1998; Nugent 2004). Nations and national development, as such, did not exist in empty, linear time that was quantifiable by the state of the economy or its enabling technological milieu (Anderson 1983; Gupta 1998). Instead, polities are situated heterochronically, partly formed but not determined by infrastructures and governmental technologies they seek to proliferate (Chatterjee 2004).

Cities and the Government of Difference

While infrastructure itself has not always been a central analytic in the social sciences, systems and norms of distribution have long interested archaeologists, historians, anthropologists, and geographers. Distribution, of course, points indirectly to the ways in which infrastructures—roads, energy networks, and water systems—redistribute resources, form polities, and have political effects. Scholars of irrigation infrastructures, like dams and canals, for instance, have demonstrated how these works, while being constructed, displace millions of residents in order to redistribute resources to a relatively more powerful few (Kothari and Bhartari 1984). This approach has been especially well developed in urban geography, where scholars have built on Marxist approaches to the built environment, focusing on the production and differentiation of space, often in direct relation to capital. Infrastructure development, Colin McFarlane and Jonathan Rutherford (2008) point out, is fundamentally a political process. Infrastructure, like science, is "politics pursued by other means" (Latour 2012: 38). Stephen Graham and Simon Marvin (2001) take a Lefebvrian (1991) approach to urban space and infrastructure. In *Splintering Urbanism* (2001), they question the singularity, ubiquity, and taken-for-granted forms of infrastructures, urging us instead to attend to them as dynamic and congealed processes of organizing finance, knowledge, and power:

> A critical focus on networked infrastructure—transportation, telecommunications, energy, water, and streets—offers up a powerful and dynamic way of seeing contemporary cities and urban regions.... When our analytical focus centers on how the wires, ducts, tunnels, conduits, streets, highways and technical networks that interlace and infuse cities are constructed and used, modern urbanism emerges as an extraordinarily complex and dynamic socio-technical process.... As capital that is literally "sunk" and embedded within and between the fabric of cities, [urban infrastructures] represent long-term accumulations of finance, technology, know-how, and organizational and geopolitical power. (Graham and Marvin 2001: 8)

For Graham and Marvin, infrastructure is an assembly of sociotechnics, and cities are made through assemblies of infrastructure. But rather than thinking of distanced or aloof hardware networks, they invoke Raymond Williams's (1973) "structures of feeling" to note that infrastructures also give shape to and are shaped by quotidian human experiences and sentiments of hope, inclusion, violence, and abandonment.

To paraphrase Susan Leigh Star (1999), to study a city and neglect its sewers and power supplies, you miss not only essential aspects of distributional justice and planning power, but also dreams and aspirations, breakdowns and suspensions, and the intimate rhythms of how we wash or go to the bathroom, how we see in the dark or cool our food, and how we travel across space (379). If urban geographers have drawn attention to the material forms of infrastructure, and the ways in which they differentiate and structure urban life, anthropologists have attended more closely to the lived experience of unequal provisioning and differentiated belonging in cities (Caldeira 2000; Chu 2014; Schwenkel 2015a). City residents often push back against this differentiated belonging, making claims to social membership, political belonging, and rights to modernity in terms of infrastructure, whether imaginary, potential, or derelict (von Schnitzler 2013). Conversely, groups may identify everyday relationships with infrastructure marked by interruption, improvisation, and modification as a metonym of their marginality (Ferguson 1999; P. Harvey 2010; Anand 2017).

Consider a road infrastructure. Communities that are not connected to the nation-state by roads often see themselves as marginalized by its absence (Harvey, this volume). Inasmuch as roads are associated with development, improvement, and modernity, roads are sites of representation and aspiration (Coronil 1997; Larkin 2008). Yet while roads are desired by political subjects, they are not always used in the ways that state planners intend (Mrázek 2002). Before long, their designs are repurposed, altered, and populated by the heterogeneous dreams, desires, and practices that confound the goals and intentions of their designers (Scott 1998; de Certeau 2002; Mrázek 2002).

Critically, while anthropologists have been especially attentive to the heterodox lives formed by infrastructure in cities, they have also drawn attention to the flexible, provisional ways in which social networks step in when material infrastructures fail to deliver (Simone 2004; Elyachar 2010). Water pipes, electricity grids, and roads are always breaking down, need constant maintenance, and are regularly being claimed by groups authorized and unauthorized by city government. Moreover, marginalized others constantly make claims on and form infrastructures beyond those controlled by the state. In these ways, infrastructures are fundamentally social assemblies (Schwenkel, this volume). In insisting we see "people as infrastructure," AbdouMaliq Simone (2004) draws attention to the ways in which social relations are a central, hidden, and vital support system necessary to live in cities. In what appears to some as the ruins of inner-city Johannesburg, a "highly urbanized social infrastructure" (Simone 2004: 407) enables people to improvise socioeconomic links with one another, providing what failed public services or formal-sector employment has not. To quote Simone, "Infrastructure is commonly understood in physical terms, as reticulated systems of highways, pipes, wires, or cables.... By contrast, I wish to extend the notion of infrastructure directly to people's activities in the city" (407). Elsewhere, Simone (2012) points out that the reticulated systems (on which this volume focuses) are themselves loci of people's activities, and they cannot be so easily dismissed as "merely" physical. As city employees and residents alike invest labor and care into everyday practices of maintenance and repair, they make more-than-human assemblies of infrastructure that are generative of differentiated materializations of rights, resources, and aspirations in the city.

Science and Technology Studies: Engineering Politics

Historians and sociologists of technology have been at the forefront of a larger turn toward infrastructure in the social sciences. In their influential work, Susan Leigh Star and Karen Ruhleder insist that "infrastructure appears only as a relational property not as a thing stripped of use" (1996: 113). Rather than being a singular thing, infrastructure is instead an articulation of materialities with institutional actors, legal regimes, policies, and knowledge practices that is constantly in formation across space and time (Ribeiro 1994; Mitchell 2002; Edwards 2003). Accordingly, infrastructures are seldom built by system builders from scratch. They are instead brought into being through compromised, improved projects of maintenance and repair (Mitchell 2002; Graham and Thrift 2007; Jackson 2014). They have histories and "grow" incrementally in a dynamic temporal, spatial, and political environment (Edwards et al. 2009: 369). They are formed with the moralities and materials of the time and political moment in which they are situated (Hughes 1983). Equatorial Guinea's national highway system and rapidly constructed gleaming buildings, for example, are very much of and in the time of oil extraction; they were built to announce the spectacular, if temporary, wealth of the petro-state to domestic residents and international visitors alike (Coronil 1997; Apter 2005; Limbert 2010; Appel 2012a, 2012b).

Emerging from the recognition that infrastructures grow temporally and incrementally, STS scholars have been particularly attentive to the emergent nature of expertise of those who manage, maintain, and extend infrastructures. Rather than assume that experts (corporate engineers, state officials, plumbers) possess an already formed expertise that they deploy to act and repair infrastructural problems, scholars have demonstrated how expertise and authority emerge from improvised, compromised heterogeneous practices that are performed amid partial knowledges and intransigent materialities (Law 1987; Bowker 1994; Harvey and Knox 2015). Infrastructures have no heroes or obvious system builders sitting in air-conditioned offices who bring them into being from a distance (Furlong 2014). Experts, often at their own admission, only have a partial knowledge of their working and are constantly compromised by the materialities and contingencies of infrastructure projects (Bowker 1994; Harvey and Knox 2015). Engineering expertise is made in the field, through efforts to repair and make infrastructures work again.

As such, sociologists of technology have attended to the labor of managing and maintaining infrastructure. Just as anthropologists might better apprehend the workings of the state by attending to the practices of lower-level governmental officials (Gupta 1998; Sharma 2008), Susan Leigh Star has urged scholars to focus on the very ordinary infrastructure workers, such as janitors and cleaners, who are otherwise unnoticed in everyday life (Star 1999: 386). These workers are vital to the everyday distributions and social life of infrastructure. Anthropologists have begun to follow Star's provocation. An attention to the practices of low- and mid-level administrators and technicians challenges any easy characterizations of technopolitics as exercised from afar (Anand 2017). Finally, STS scholars have also urged scholars of infrastructure to pay more attention to "those at the 'receiving' end of infrastructure"—those who are subjected to its distribution regimes and marginalizations in everyday life (Edwards et al. 2009: 371). As Edwards et al. ask, "How can claims on, through, and against infrastructure be formulated, organized, and heard? What constitutes adequate representation or participation in the process of infrastructural change and development? Under what conditions can rival interests in infrastructure (large and small, modest and profound) be acknowledged, addressed, and accommodated, in ways that enhance the legitimacy, appropriateness, and long-term efficacy of infrastructural change?" (Edwards et al. 2009: 372).

Anthropologists are well positioned to answer these questions through ethnographic studies of infrastructure. An attention to the materialities and socialities that are gathered to form infrastructures promise to both demonstrate how these vital support systems are formed, and also how they bring other things into being and constitute social worlds (Ferguson 2012: 559).

In this volume, we draw together insights from STS, urban studies, and development studies to ask critical questions about how anthropologists study infrastructure. What happens when infrastructure is no longer a metaphor? What happens to theory making and ethnographic practice when roads, water pipes, bridges, and fiber-optic cables themselves are our objects of engagement? How do we take seriously the developmentalist fantasies and desires for modern infrastructure, often articulated by marginalized subjects themselves (Ferguson 1999)? A focus on infrastructure enables us to consider seriously the articulation (and disruption) between the technologies of politics and the politics of technology (Barry 2001; Anand 2011). By shifting our attention to infrastructures as ethnographic objects we promise new theoretical and political insight both for anthropology and from anthropology. Like Chakrabarty's (2000) provocation to "provincialise Europe," attention to infrastructure forces us to ask: What do we see differently and understand otherwise when we shift the analytic center? We see studies of infrastructure as a forceful reengagement with gender, race, colonialism, postcoloniality, and class on new empirical and political terrain. Infrastructure provides a site in which these forms of power and inequality are reproduced or destabilized, in which they are given form, occasionally obduracy, and often contingency. Precisely because all knowledge is situated (Haraway 1991), where we think from and what we think about affects what it is that we are able to think. Thinking from and with infrastructure allows new and productive decenterings and provincializations: space-time compression, for instance, depends on contested material and aesthetic choices; and liberal governance is no longer rationality at a distance, but politically intimate practices. And indeed these are the themes—time, politics, and promise—around which we organize the volume and the remainder of the introduction.

Time and Temporality

Brian Larkin (2013) begins his influential essay on infrastructure by observing that "infrastructures are built networks that facilitate the flow of goods, people, or ideas and allow for their exchange over space" (328). Space and spatiality naturally come to mind when thinking of infrastructure's features and effects. Infrastructures bridge distance; roads, railways, wires, and pipes help connect one point to another, and they are heavily dependent on and constitutive of local geographic contexts (Hughes 1983; Coutard et al. 2004).

Whether one thinks of channels of transport like railway lines or flight paths, of electric and communication wires, or the movement of resources like water or oil in pipelines, it is the connection through space that is central to the working of infrastructure. As Larkin observes, "For some time now, scholars in science and technology studies and geography have analyzed how infrastructures mediate exchange over distance, bringing different people, objects, and spaces into interaction and forming the base on which to operate modern economic and social systems" (2013: 330). But infrastructure, of course, mediates time as much as it mediates space (Degani 2013; Hetherington 2014). Infrastructures configure time, enable certain kinds of social time while disabling others, and make some temporalities possible while foreclosing alternatives (Barak 2013).

Revisiting two ideas that have become commonplace in the present—time-space compression and just-in-time production—demonstrates the imbrication of time with infrastructure with particular clarity. First, the notion of space-time compression, popularized by David Harvey in *The Condition of Postmodernity* (1989), referred in part to the ability to conduct "real-time" financial transactions across the globe. Second, just-in-time production referred to supply-chain changes enabled by global shipping, containerization, and the global factory floor, all of which sharply cut the capital tied up in products sitting in warehouses by more closely aligning supply with demand, thus significantly reducing the chances of a crisis of realization (D. Harvey 1989). What were the infrastructural conditions that facilitated the move toward real-time transactions? Much of this story hinges on an effort to lay undersea fiber-optic cables that connected global financial centers. As Nicole Starosielski's (2015) work demonstrates, this messy project is deeply entangled with daily life in Guam, undersea aquatic life, and both colonial and cold war telegraph and telephone infrastructures. Just-in-time production too depended on the installation of communications infrastructure, as well as new technologies for inventory control and management, not to mention containerization and global shipping (Carse 2014). These infrastructures of contemporary capitalism were developed over long periods of time, in a process that was neither linearly progressive nor uniform (Rosenberg 1976; Elster 1983). Thus, time-space compression is itself a temporal process that comes into being with the simultaneous development of new technologies of communication (fax machines, fiber-optic cables), a massive investment of capital and labor to connect vast distances with these technologies, and new methods of managing inventories and logistics (Sassen 1991; Starosielski 2015). Once installed, these infrastructures introduced new (and always more fitful than portrayed) temporalities in the worlds of finance, commodities, and labor, which in turn changed the nature and experience of social time and social space.

In *Knowing Capitalism* (2005), Nigel Thrift writes that "it is all too easy to depict capitalism as a kind of big dipper, all thrills and spills. But capitalism can be performative only because of the many means of producing stable repetition which are now available to it and constitute its routine base" (3; see also Gupta 1992). Infrastructure, of course, is chief among these means of producing the more or less stable performance of both real-time transactions and just-in-time production. While Thrift does not use the term infrastructure specifically, he seems to be urging our attention to it—toward what he calls "the apparatus of installation, maintenance, and repair" on the one hand, and "the apparatus of order and delivery" on the other." Thrift writes, "For some reason, perhaps to do with their extreme everydayness, these apparatuses are constantly ignored in the literature, and yet it could be argued that they constitute the bedrock of modern capitalism" (2005: 3).

Thinking of time-space compression and just-in-time production through infrastructure paradoxically draws attention to the slowness of the process of speeding up. For example, it draws our attention to the resistance present in telegraph or fiber-optic cables—the difficulties involved in financing them, and then in installing and repairing them (Barak 2009; Starosielski 2015). Rigorous attention to infrastructure itself actually slows time-space compression enough to see delay, accretion, suspension, repair, resistance, and repurposing. Ethnographic attention to infrastructure may "ultimately undermine any idea that speed or time economy—the grossest simplification of efficiency's logics—is at the heart of capitalism. Instead, we will be able to explore the heterogeneous forms of pacing, duration, waiting, pause, obsolescence, and delay that also characterize its generative rhythms" (Bear et al. 2015: sec. 8; see also Bear 2015).

The larger point here is not to displace thinking about space by the logic of time, nor to privilege time over space. Focusing on time and temporality, in fact, helps us think of spatiality in new and interesting ways; it allows for a re-thinking of spatialization as itself a temporal act and activity (Althusser 1969; D. Harvey 2005; Gidwani 2008). Temporality is built into spatial expansion, contraction, and scaling. Attention to the life span of infrastructure itself slows us in this way, hence attuning us to the shifting social temporalities that infrastructure produces. For example, a new metro system is rarely put into place all at once (Latour 1996). Instead, one main line is first prepared and started, which changes the time that it takes for people to commute from their homes to workplaces, or from one part of the city to another. It also changes the urban form, as new housing, offices, and shops spring up along the metro line. Real estate markets shift as valuations of property that might have formerly been thought undesirable go up because they now are adjacent to the metro line. Then, as new lines are added to the metro, further shifts take place in commute times and in the urban form. The temporality of infrastructure, therefore, matters a great deal in the creation of spatial patterns of living, working, and entertainment; it influences the direction and degree of spatial extension; and it has profound social and political impacts. Like metros and rail lines, highways, cable networks, and even wireless communication all extend spatially over time, and it is this temporality that in turn produces variegated forms of spatiality and particular patterns of sociality.

A perfect example of the temporal interplay between spatial and social extension is provided by Antina von Schnitzler's work on metered water and electricity in South Africa. In response to the supply of expensive water and electricity in the townships of South Africa, residents found a way to tamper with the meter so that they did not have to pay high bills. In response, utilities began providing connections through a new tamper-proof technology of prepaid meters. In this case, the spatial extension of piped water and electricity encountered political resistance, but precisely because the spatial extension unfolded over time, that resistance was itself enfolded in a new technology of governmentality. The spatial, the temporal, and the political were mutually produced in this encounter. Or consider Nikhil Anand's work on water politics in Mumbai: the spatial extension of pipes that can potentially supply nonauthorized settlements with municipal water depends on a politico-temporal trajectory, one in which elaborate negotiations occur over an extended period of time among slum residents, politicians, and engineers and managers in the hydraulic bureaucracy. The outcome of such negotiations is always uncertain and subject to revision. Thus, the relationship between having pipes and having water is always up for grabs, and it can swing one way or the other depending on the social and political climate.

Once we conceptualize infrastructures not just in terms of the different places that they connect, but as spatiotemporal projects—as chronotopes— then we can open up new ways of thinking about the temporality and spatiality of infrastructure. As opposed to the "finished" product of a planner's map, if we think of infrastructures as unfolding over many different moments with uneven temporalities, we get a picture in which the social and political are as important as the technical and logistical (Gupta, this volume). Another way to say this is that conceptualizing infrastructure as a process over time ensures that the technical and logistical sides of infrastructure are not privileged over, or seen as separate from, its social and political, or formal and aesthetic sides

(Larkin, this volume; Schwenkel, this volume). Paying attention to the temporality of infrastructure makes us aware that the same technical features can produce very different configurations of space and sociality than those designed by planners. Many projects do not work out as planned because, as they are implemented, social and political pressures force alterations in their design and in their function.

A processual view of infrastructure focuses on infrastructure's protean forms (Star and Ruhleder 1996; Graham 2010). Looking both across and even within the different phases of infrastructure's life span—design, financing, construction, completion, maintenance, repair, breakdown, obsolescence, ruin—one can see the operation of multiple temporalities and trajectories. For example, as Gupta's essay in this volume points out, once an infrastructure project is started, it does not necessarily have to be completed. It can be suspended or abandoned, delayed or deferred. Abandonment and suspension can result from social and political struggles regarding the project, or they can be an outcome of technical, political, and financial failures.

Even after a project is completed, it is always changing. These changes are due to the materiality of the infrastructure itself. Decay and deterioration affect all materials. For instance, pipes made of steel, copper, or PVC will have different rates and probabilities of failure over time. The life cycle of materials may create high or low probabilities of breakdowns and ruptures. And yet, however important materials may be for explaining failures, referring to the qualities of materials in isolation is insufficient. Time and infrastructural life spans are made relevant through historical relations with others.

We also have to take into account gaps in knowledge or a lack of resources for routine maintenance. More importantly, the social and political life of infrastructure changes over time. With the rise of air travel, railway stations decline in importance. Similarly, gas stations may become less ubiquitous with the move to electric cars. Highways and metro lines often split existing communities or are used in processes of gentrification to displace certain residents and welcome others (Winner 1999; Graham and Marvin 2001). And what happens when highways or train lines are discontinued? When infrastructures seem to disappear? In an essay originally prepared for the project from which this volume emerged, Catherine Fennell asks what new kinds of sociality and obligation come into being when infrastructure is abandoned (Fennell n.d.)? Her essay on house demolition in the United States' late industrial Midwest underlines the point that abandonment is not a moment but a process. As urban housing stock is razed, political controversies and public health concerns rise alongside dusts released through the destruction of cities considered "overbuilt" in the wake of deindustrialization. How is the construction and reproduction of social time altered as a result? A focus on the temporal helps us think of the spatial, technical, material, logistical, political, and social properties of infrastructure together.

Focusing on temporality, however, is important for another reason that has to do with what Larkin (2013) has called the poetics of infrastructure. Infrastructures are important not just for what they do in the here and now, but for what they signify about the future. Particular infrastructures signal the desires, hopes, and aspirations of a society, or of its leaders. Nation-states often build infrastructures not to meet felt needs, but because those infrastructures signify that the nation-state is advanced and modern (Ferguson 1999; Apter 2005; Appel 2012b; Harvey and Knox 2015; Gupta, this volume). That is why there is always greater investment in future-oriented infrastructures than is justified by their expense. Shiny new airports with huge capacities are built in many countries although they only serve a tiny elite, whereas less glamorous infrastructures, which would actually be more useful to the poorer segments of the population, are ignored or overlooked. This aspiration toward a modern future is often widely shared by large segments of the population. Focusing on these investments makes evident a hitherto hidden temporality. Different visions of the future, different aspirations for one's own life, and for the future of the community or nation, play an important role in shaping which infrastructure projects find support among populations. Like the building of infrastructure itself, emotional and affective investments, too, are not forged once and for all, but take time to be formed, and can change from positive to negative over a period of time. In other words, the affective relationship between people and infrastructure, while being shaped by notions of futurity, is not always positive and may instead result in deferral, ruination, suspension, abandonment, and repurposing (Stoler 2013).

Finally, there is another way in which the temporality of infrastructure proves to be important. Infrastructures often exceed human lifetimes (Bowker 2015). Bridges, roads, and buildings live longer than most humans and promise to continue to matter after our death. Even in a relatively young nation-state such as the United States, the oldest, continuously used bridge dates to the late seventeenth century. The effects of infrastructures dependent on fossil fuels—from coal-fired power plants to superhighways, flaring oil and gas wells to airline travel—will persist in the atmosphere for many generations. Nuclear wastes will probably pollute the earth long after humans are extinct. Global oil infrastructure offers a good illustration of the long lifetime of infrastructure, both in its own tenaciously obdurate life span and in its afterlives. A sketch of this massive infrastructural network would include 5 million producing wells across the world (3,300 of which are subsea); more than 2 million kilometers of surface pipelines and another 75,000 kilometers running along the seafloor; 6,000 fixed oil platforms and 635 offshore drilling rigs; and more than 4,000 oil tankers moving 2.42 billion tons of oil and oil products every year (Bridge and Le Billon 2013; Appel, Mason, and Watts 2015). This is a radically work-intensive and dynamic infrastructural network, constantly under repair, maintenance, expansion, and contraction as certain resources are exhausted and others are accessed, as geopolitics change, and as cost-effectiveness and political viability of various extractive technologies evolve. The network is also unevenly defined not only by exploration, extraction, transport, and delivery, but also by leakage, breakage, and sabotage. The Anthropocene, intimately bound to fossil fuels, focuses our attention on the temporalities of this network—not only the grave inertia in the face of needed retrofit or conversion to other fuel sources, but also, of course, the atmospheric afterlives of all its combusted products that will long outlive the wells, pipelines, and tankers themselves. Thinking of the temporality of infrastructure draws us into another way of thinking beyond the human that has to do with other timescales, times that are not scaled (down) to human life, and only draw meaning from those lives. To decenter humans is in part to think about other time spans, the lifetimes of other things that shape life on the planet, and infrastructure is one important element in such a rethinking.

It is perhaps a sign of infrastructure's capaciousness, however, that we can use it to think across radically different scales—from planetary epochs to the most intimate acts of individuals, families, and communities. It is to these questions of intimacy—what Fennell has described as infrastructure "pressing into the flesh"—that we now turn to look at how infrastructure allows us to reframe politics.

Infra Politics: Intimacy, Publics, Matter

Politics

Infrastructures are a critical site through which politics is translated from a rationality to a practice, in all its social, material, and political complexity (Humphrey 2005). They are a material and aspirational terrain for negotiating the promises and ethics of political authority, and the making and unmaking of political subjects. Because infrastructures distribute vital resources people need to live—energy, water, information, food—they often become sites for active negotiations between state agencies and the populations they unevenly govern. As such, several authors in this volume theorize how politics is enacted in everyday life by attending to the performances of infrastructure. To do this, they suggest we attend more closely to infrastructural forms, the ways in which their administrations gather publics and distribute life not just in terms of aspirations and possibilities, but, in their very material sense, to flourish and to proliferate life. After all, life not only has meaning, and is not just symbolic, but is also matter. Life's materiality, Didier Fassin reminds us,

> is not simply, in the Marxian sense, that of the structural conditions which effectively largely determine the conditions of life of the members of a given society; it is also, in Canguilhem's sense, that of the very substance of existence, its materiality its longevity and the inequalities that society imposes on it. To accept this materialistic orientation is not a merely theoretical issue. It is also an ethical one. It recognizes that the matter of life does matter. (2011: 193)

The #BlackLivesMatter movement and the Movement for Black Lives have argued along similar lines, where the infrastructural violence of Detroit and Flint sharpens claims to infrastructural justice and invest/divest platforms demand a "rebuilding and repair plan for domestic infrastructure across the country based on a commitment to a green economy and deep understanding of the threat of climate change" (https://policy.m4bl.org/invest-divest/). Fassin, in other words, is not alone in gesturing to the politics of distribution in recent years (see also Ferguson 2015). The Foucauldian scholar Thomas Lemke (2014) recently urged "relational materialism" for exploring the emergence of liberal modes of rule through material forms. To understand its processes, Lemke suggests, we need to articulate more carefully the link between the "matter of government and the government of matter" (2014: 14). To govern infrastructure, we argue, is to govern the politics of life, with all its inequalities.

Infrastructures are critical sites for the distribution of life and a key locus for the performance (and theorization) of politics and polities today. For instance, consider the ways in which water infrastructures are designed, installed, and managed in cities. To produce a hydraulic infrastructure does indeed require a population to be defined, delimited, and imagined through a planning process. Throughout this process, government officials, hydraulic engineers, and construction firms are called upon to design a system that can serve a given population. This population is itself discursively formed through practices of counting and measurement. Nevertheless, this population is only socially and materially brought into being with and through the everyday flow of water through the pipes that (do or do not) ensue, a population that Anand (this volume) has termed "hydraulic publics." Alignments of the pipes, politics, laws, and policies materially gather, constitute, and manage the population of the city. As a governmental service, or a privatization controversy (as in Detroit, Delhi, or Mumbai), water pipes define not only who its subjects are, but also how they are collectively and differentially "treated" by the (public or private) institutions that administer infrastructures. Through everyday connections and disconnections, pipes, roads or electricity wires form populations that are unevenly governed and left aside.

This kind of attention to infrastructure unsettles long-accepted understandings of how rule is accomplished. If "liberalism is a form of government that disavows itself, seeking to organize populations and territories through technological domains that seem far removed from formal political institutions" (Larkin 2013: 328), ethnographic attention to infrastructure—from public toilets to municipal water systems, from roads to leaded homes—forces us to rethink governance and citizenship not at a distance but pressing into the flesh, through questions of intimacy and proximity. Taken together, the authors in this volume describe not only how infrastructures interpellate specific types of subjects, but how the immediacy and intimacy of infrastructures enable those subjects to "hail the state" and recognize them as publics (Anand, this volume). Infrastructure does not allow state power to disavow itself. On the contrary, it is an intimate form of contact, presence, and potential, one that serves as an important locus for the evaluation of the morality and ethics of political leaders and the state.

Indeed, the very idea of "disruption" operates with the assumption that quietly working infrastructures are "normal." A "disruption" such as a strike or work slowdown, draws attention to all the labor that goes into making infrastructures work invisibly and quietly behind the scenes. Thus, while infrastructures may be planned or designed with certain populations in mind, Christina Schwenkel demonstrates how these are frequently repurposed by others through political claims and material practices. We are arguing neither that the socialities and polities of populations are effects of infrastructural forms, nor that infrastructural forms just result from human social and political relationships and technologies (see Larkin, this volume). Instead, as infrastructures such as roads, communications infrastructures, or energy grids are built over more-than-human lifetimes (Braun 2005; Bowker 2015), populations and publics are constituted iteratively—with and through the materials of infrastructure, not just prior to or following its construction (Braun and Whatmore 2011).

Publics, therefore, are not only formed and controlled through an extension of infrastructures. By tugging, pulling, and demanding infrastructures to recognize, serve, and subjectify them, publics also make themselves visible as demanding subjects of state care. This is true, of course, not only of public or service delivery infrastructures, but of private infrastructures as well. To the extent that infrastructures of production, transport, delivery, and marketization enable the daily performances of capitalism, they are also central to potential disruptions (Mitchell 2011). Strikes, roadblocks, port shutdowns, work slowdowns, and sabotage all hinge on access to infrastructure in order to disrupt the exchange of goods, people, resources, and ideas over space (von Schnitzler 2016).

Like the building of a road, the constitution of populations is a project: a work-intensive endeavor that is often sporadic and stuttering. Infrastructures do not just help us to rethink politics, offering new ground on which to watch how publics take shape and disperse. They also help us think through the classical literature on publics (Habermas 1989; Warner 2002) to show how publics are not just made through enunciatory communities, circulations of intention, text, and speech that produce disembodied spheres of deliberation and fantasies of free circulation. The authors in this volume show how publics are also made by infrastructures that assemble collectives, constitute political subjects, and generate social aspirations. Different chapters in this volume suggest how publics are both brought into being and forestalled through infrastructural processes, such as the construction of a highway (Harvey, this volume), shared defense of bridges and power plants in bombarded cities (Schwenkel, this volume), the politics of municipal service delivery (von Schnitzler, this volume), or the mass destruction of urban housing stock in the U.S. Midwest (Fennell n.d.).

Of course, there is no predictable relation between infrastructures and the production of publics. As von Schnitzler (this volume) points out, water meters in South Africa have the capacity both to constitute publics and to truncate or fracture them. In other words, the fact that infrastructures can participate in the making of publics does not mean that they always do so. On the contrary, the vitality of their material form is overdetermined, and infrastructures can just as easily impede and proscribe the formation of publics (Rodgers and O'Neill 2012; Harvey and Knox 2015). Appel (2012b) shows that a surfeit of private infrastructure investment by U.S. oil firms in Equatorial Guinea, reminiscent of colonial infrastructures, works to provision markets and expatriates while deliberately excluding Equatoguineans, so that the oil companies are not held responsible for Equatoguineans or liable for any harm done to the population. Outside the walls of U.S. oil enclaves, Equatoguinean state representatives routinely attribute a deferral of "the public"—the tenacious problems of electricity distribution, a potable water supply, healthcare and education provisioning, not to mention the draconian limits on press and political organizing—to the need to focus first on infrastructure. Here, materialized infrastructures are both "objects of the future and justifications of the present's constant deferral" (Appel, this volume).

Matter

That publics can be gathered or forestalled by the materials of infrastructure, often despite the political imaginaries or social aspirations embedded in their design, complicates any neat relation between matter, human-centered accounts of politics, and infrastructural form. In recent years, new materialist scholarship, drawing on more established research in the STS field, has urged that nonhumans be included as actors in the theorization of politics (Latour 1996; Bennett 2010). For the new materialists here, nonhumans could include other life forms—like scallops (Callon 1984); jaguars (Kohn 2013); mountains, or other types of matter perceived to be inanimate in Christian cosmologies (de la Cadena 2010); or mechanical coupling systems (Latour 1996). Nonhumans, vital materialists remind us, are "vibrant" and have both an existence and a "force" that is "beyond the human" (Bennett 2011; Kohn 2013; Meehan 2014). Such scholars urge us to attribute some form of political agency to different kinds of actors in an infrastructural assemblage. Yet, as Arjun Appadurai (2015) has pointed out, new materialist scholarship continues to lack a sufficient accounting of history, difference, and responsibility (O'Neill and Rodgers 2012). As Appadurai asks provocatively,

> If agency in all its forms is democratically distributed to all sorts of dividuals, some of which may temporarily be assembled as humans and others as machines, animals, or other quasi agents, then do we need to permanently bracket all forms of intrahuman judgment, accountability, and ethical discourse? Will future courts only be judges of assemblages of hands-guns-bodies-bullets and blood or of syringes-heroin-junkies-dealers or of ricin-envelopes-mailboxes-couriers and the like? And, worse, who will be the judges, witnesses, juries, prosecutors, and defenders? Will our very ideas of crime and punishment disappear into a bewildering landscape of actants, assemblages, and machines? If the only sociology left is the sociology of association, then will the only guilt left be guilt by association? (2015: 24)

Several essays in this volume have engaged these contemporary debates, as we try to widen the field of political action while staying carefully tuned to our ethnographic research that reveals historical and social accretions of difference, marginality, and responsibility between human actors. Together, we demonstrate how matter and nonhuman relations are not just an inert substrate that yields to the dreams and desires of powerful (human) actors. Instead, infrastructure's materials are active participants in its form and, therefore, also its politics. For instance, in this volume Dominic Boyer argues, via a careful reading of Marx, that we might think of infrastructure's materials as "a potential energy-storage system, as a means for gathering and holding productive powers [of capital, but also of state] in technological suspension." As a congealed and transformed process of people and things, infrastructure is generative and transcendent. Yet, its effects are themselves heavily contingent on the mediations of its materials. A misalignment of these material forms can, as we noted in the case of the Deepwater Horizon spill, "blow the very same [tenuous] arrangement 'sky-high'" (Boyer, this volume).

Nevertheless, at the same time, we do not attribute to the materials of infrastructure either a primacy or an independent existence that is discrete and beyond (human) history (Larkin, this volume). Thus, Penny Harvey (2015) points out that "the point of interest is not simply that materials always carry their own vitality (Ingold 2012) or exert a degree of autonomous force or agency (Bennett 2010), but rather that this force is never generic, nor is it simply material." Concrete, steel, copper, and the other materials of infrastructure are historical forms that emerge through and with social systems of ideology, meaning, and imagination. They are simultaneously technical and aesthetic devices that are brought into being by relations between imagination, ideology, and technicity that have a history, a present, and a future, constantly "producing new experiences of the world" (Larkin 2015). In this volume, we refuse to accord primacy either to the powers of human representation to account for material forms, or to the powers of materials in their imagined, ahistorical, elemental state to determine infrastructural forms. Instead, as things become political only through relations (Barry 2001), we call for a recognition that materials and ideology together participate in the makings of infrastructure, politics, and publics. These relations are uneasy and difficult to anticipate in advance but can be revealed through careful historical and ethnographic research.

If material is always caught up in meaning, an attention to the materiality of infrastructure reveals how it is central to the sensory, somatic, and affective ways in which we inhabit this world (Mrázek 2002; Larkin 2013). On one hand, we notice, and make meaning of, infrastructure through a variety of senses. As

Schwenkel argues, "Infrastructure, broken or not, often evokes a multiplicity of embodied sensations across the human sensorium" (2015b). Infrastructures excite affects and sentiment, whether it is people hearing the roar of a highway, or feeling the stillness of time and air on their skin during an electricity blackout, or seeing a particular kind of past in the abandoned chimneys of a socialist energy plant. Infrastructures produce a sense of belonging, accomplishment, or loss, as polities are constantly being unmade and remade through not only the things that infrastructures carry, but also the semiotic and sensory ways in which they shape being.

But infrastructures are not just sensed; they also "press into the flesh," materializing and "emplacing" new figurations of the body (Fennell n.d.). The mortgaged American home, Fennell underscores, is central to national projects of wealth expansion, economic mobility, and collective provisioning. Focusing on large-scale housing demolition in the late industrial, urban U.S. Midwest, Fennell highlights the "noxious aspects" of political emplacement. She tracks how those tasked with removing vacant houses grapple with the potential health effects of abandoned, decaying, but also demolished, infrastructures. Here the process of a retreating infrastructure does not just entail the disconnection of political subjects. The abandonment and demolition of the house are also experienced as a toxic uncertainty, one whose materials might become absorbed by a body and continue to inhabit it, long after the promise of infrastructures—and the dreams of access and recognition that constitute them—has been carted away as rubble.

Conclusion: On Promise

As Susan Leigh Star has noted, "One person's infrastructure is another's topic, or difficulty" (1999: 380). "Modern" infrastructure may be as various as a contractual obligation between transnational oil corporations and states (Mitchell 2011; Appel 2012b), or a set of high-speed fiber-optic cables that, viewed from one perspective, allow twenty-four-hour workdays, but, from another, disrupt the sleep and kinship patterns of call-center workers in Bangalore (Aneesh 2006; Starosielski 2015). Whether they are being built or crumbling, infrastructures simultaneously index the achievements and limits, expectations and failures, of modernity. We inhabit worlds already formed by differentiated infrastructures, making them good to think with in the classic ethnographic sense. In this volume, we see how the material and political lives of infrastructure frequently undermine narratives of technological or social progress, drawing attention instead to the shifting terrain of modernity, distribution, inclusion, and exclusion in most of the world. As materials and technologies transform, so do their promises. New infrastructures are promises made in the present about our future. Insofar as they are so often incomplete—of materials not yet fully moving to deliver their potential—they appear as ruins of a promise. Infrastructures in Equatorial Guinea, Bangalore, or Detroit do not only promise a future. Suspended in the present, they symbolize the ruins of an anticipated future, and the debris of an anticipated or experienced liberal modernity. In the final section of this introduction, and in the chapters that close the volume, we engage more explicitly with the promise of infrastructure.

At its most basic, the promise of this volume is that ethnographic attention to infrastructure offers new empirical terrain on which to understand the entanglements of social, political, economic, biological, material, aesthetic, and precarious, threatened life. Attention to the lives of infrastructures helps us think about possible worlds and new relations between life, matter, and knowledge in the Anthropocene, or what Donna Haraway (2016) has usefully expanded to the Anthropocene, Capitolocene, and Chthulucene. Infrastructures are promising locations for ethnographic research precisely because they are sites of conceptual trouble that refuse the easy separation of the human and the material; they are more-than-human forms that demand acknowledgment of sociomateriality. Infrastructures are where aesthetics, meaning, and materiality meet (Fennell n.d.). They promise an assessment of social life that "sidesteps the limits of humanism without erasing the human, and . . . allows for a dynamic and open sense of scale that does not assume a singular perspective" (Harvey, this volume).

Each chapter in the volume thinks through the question of promise not only as the intellectual potential of the task at hand, but also as an embodied experience of infrastructure itself around the world. Opening the book's section devoted to time, Hannah Appel writes about the lived experience of spectacular rates of infrastructural investment and construction in Equatorial Guinea. Noting infrastructure's centrality to ideas of the national economy and economic growth, Appel argues that attention to infrastructure lays bare the promises and betrayals of developmental time, oil time, political time, and imperial time. Infrastructural time emerges as a new mode of attention to archaeologies of the present, affording multiscalar insights into today's imperial formations and the poisoned promise of economic growth.

Akhil Gupta continues these archaeologies of the present in chapter 2, with a focus on suspension and a rereading of ruins through half-built infrastructure projects in Bangalore. Rather than the elegiac decay of once-flourishing civilizations, Gupta terms the incomplete infrastructure projects "the ruins of the future," wherein "ruination is not about the fall from past glory but this property of in-between-ness, between the hopes of modernity and progress embodied in the start of construction, and the suspension of those hopes in the half-built structure." If promise is "that which affords a ground of expectation of something to come," Penny Harvey attends to the longing many Peruvians articulate in relation to roads in chapter 3. In a moment when infrastructures seem to be the current currency of investment across much of the global South, she asks, what are the grounds of expectation that accompany infrastructural projects? Concluding the section on time, Christina Schwenkel shows how smokestacks in postcolonial Vietnam concretized the promise of technological prosperity, but were always already alloyed not only with risk and calamity, but also by political subjection. Resonant with Larkin's attention to infrastructure's political aesthetics in chapter 7, Schwenkel shows us in chapter 4 that under the banner "Đảng là ánh sáng" (the Party is the light), electrification projects brought new symbolic legitimacy to the postcolonial government at the same time as they justified novel forms of oppression and desperation.

Antina von Schnitzler's work opens the book's second section on infrapolitics. Through the unlikely ethnographic subject of the toilet, von Schnitzler argues in chapter 5 that infrastructure's epistemological promise is not given but contingent. She writes that, for her purposes, "infrastructure is both an ethnographic object and an epistemological vantage point from which to understand a less apparent postapartheid political terrain and a location from which the South African present may be defamiliarized and the political rethought." Indeed, the promise of the book's second section is that ethnographic attention to infrastructure yields new insights with which to theorize politics and political life. As Nikhil Anand shows in chapter 6, infrastructures (or their absence) may not only constitute and control populations, but they are also dream worlds of promise that are actively desired and called upon by marginalized groups. Attention to infrastructure allows us to show the making and management of difference—class, race, gender, religion, and beyond—in the technics and politics of everyday life.

Together, the chapters in the second section join the domestic home to the apparatus of the state, and less frequently to capital, showing infrastructures as sites of conceptual and scalar trouble. For this reason, they offer a creative location for the production of political theory. For example, infrastructures often quite literally connect and constitute boundaries between public and private, boundaries that people sometimes reject or attempt to transgress. Governance, it turns out, does not take place at a distance but through the intimacy and proximity of toilets, pipes, and potholed roads.

Brian Larkin begins the book's final section on the question of promise. Thinking through form and political aesthetics, Larkin argues in chapter 7 that infrastructures "address the people who use them, stimulating emotions of hope and pessimism, nostalgia and desire, frustration and anger, that constitute promise (and its failure) as an emotive and political force." Larkin uses the idea of promise to critique the suggestion that matter can be apprehended apart from or a priori to form. "The promise of infrastructure," he writes, "derives from exactly the political rationalities, sense of expectation, and desire that take us into the realm of discursive meaning.... The very word 'promise' implies that a technological system is the aftereffect of expectation; it cannot be theorized or understood outside of the political orders that predate it and bring it into existence." To think about infrastructures productively, Geoff Bowker insists in the chapter that follows, is also to acknowledge the problematic intellectual orders in which most of us have been trained. As Bowker argues in chapter 8, the Baconian classification of academic disciplines, produced in the eighteenth and nineteenth centuries, specifically sought to distinguish between the human and natural sciences and between technology and society. Produced in a historical moment where the sciences were being figured by Enlightenment thinkers, these divisions continue to constrain academic labor. They constrain and shape—enframe—the study of infrastructure, an inherently transdisciplinary (or de-disciplinary) phenomenon. If we are living in the ruins of Cartesian dualism, then we will need not only new material infrastructures, but also new epistemological infrastructures to confront the present moment. Bowker asks, "How do we reimagine the nature of knowledge for the way the world is now? How do we put into infrastructures forms of knowledge production that can bear the weight of these new exigencies?"

In the chapter that concludes the volume, Dominic Boyer is also concerned with the relationship between infrastructure and epistemic and political orders, and his question is about the revolutionary dissolution of our current order, and infrastructure's potential role therein. Boyer concludes by asking how infrastructures themselves shape what is possible to think about energy use and climate change (see also Jensen n.d.). The "impossibility" of a rapid shift to renewable energy is often articulated around sedimented infrastructures like pipelines or electrical grids. And yet, if infrastructures distribute power, they are also sites of vulnerability (Mitchell 2011). The same arrangements that produce their dominance can (and perhaps are) being blown sky-high with the climate crisis. As nation-states seek to proliferate renewable energy sources following the Paris climate accord, wind and solar infrastructures are generative of new forms of sociality and may revolutionize hierarchies and privileges accreted in existing distribution regimes. Through a novel reading of Marx, Boyer encourages us to look at infrastructure as potential, radical energy; he asks how infrastructure can do or "promise" something other than to reenable the present anthropocentric trajectory. Drawing on Sara Ahmed's (2014) work, Boyer intertwines the call for revolutionary infrastructure with a call for a revolution in "transcendence-seeking 'hypersubjects' (usually but not exclusively white, straight, northern males) that gifted the world the Anthropocene as part of their centuries-long project of remaking the planet for their own convenience and luxury." He quotes Ahmed (2014) who wrote, "It takes conscious willed and willful effort not to reproduce an inheritance," and fittingly ends with a plea: please do what you can to help.

The concluding chapters on the theme of promise, then, show infrastructure as not just theoretically generative, but also essential to contemplate in the political and environmental conjuncture we are writing and thinking in. If infrastructures of energy and other resources have been key in enabling the sustaining of unsustainable rates of resource extraction, energy consumption, climate change, and planetary transformations that are now being called the Anthropocene, it is in their re-forming and re-making that we might imagine and produce different relations of distribution and circulation.

What kinds of infrastructure—epistemic, energetic, political—might we contemplate from the everyday ruins and rubble wrought by infrastructure today? How might we reimagine their forms and potentialities anew in times where the end of life itself has been rendered thinkable? In an era where federal commitment to knowledge production is weakening, and universities and institutions are struggling to maintain public funding, a politics of infrastructure necessarily asks after its potential to transform the world we currently inhabit.

Ethnographic work, we suggest, can help redeem the promise of infrastructure by making more visible, and indeed more political, the formative role of infrastructures in the ways we think, build, and inhabit our shared futures. And this may be most timely and necessary in the Anthropocene. As the promises of modernity are crumbling under neoliberal austerity and climate change, the ruins of liberalism are manifest in the sociomaterial remnants of oil wells and superhighways, water pipes and shipping channels, fiber-optic cable and an ever-growing pile of rubble. What kinds of futures and future polities will today's infrastructures leave behind? What are the dreams that may be re-gathered amid its rubble (Gordillo 2014; Jackson 2014)? When the infrastructures of history continue to reverberate in our figurations of the future, what kinds of structures and limits do they leave us with? As the infrastructures we live with are remade, they provide us with an opportunity to think, imagine, and rebuild the world differently. Can we produce a world that can be distinguished from the constitutive divisions of modernity and its progressive readings of the future, given that the epistemic and concrete glue of infrastructure binds that future to our present and our past?

REFERENCES

Act 436 of 2012. Michigan Compiled Laws Complete § 141.1541–75, *The Michigan Legislature Website.* http://legislature.mi.gov/doc.aspx?mcl-Act-436-of-2012.

Ahmed, Sara. 2014. White Men. *feministkilljoys,* November 4, 2014. https://feministkilljoys.com/2014/11/04/white-men.

Althusser, Louis. 1969. *For Marx.* Translated by B. Brewster. New York: Pantheon Books.

Anand, Nikhil. 2011. Pressure: The PoliTechnics of water supply in Mumbai. *Cultural Anthropology* 26: 542–564.

———. 2017. *Hydraulic city: Water and the infrastructures of citizenship in Mumbai.* Durham: Duke University Press.

Anderson, Benedict R. 1983. *Imagined communities: Reflections on the origin and spread of nationalism.* London: Verso.

Aneesh, A. 2006. *Virtual migration: The programming of globalization.* Durham: Duke University Press.

Appadurai, Arjun. 2015. Mediants, materiality, normativity. *Public Culture* 27(2): 221–237.

Appel, Hannah. 2012a. Offshore work: Oil, modularity, and the how of capitalism in Equatorial Guinea. *American Ethnologist* 39: 692–709.

———. 2012b. Walls and white elephants: Oil extraction, responsibility, and infrastructural violence in Equatorial Guinea. *Ethnography* 13(4): 439–465.

Appel, Hannah, Arthur Mason and Michael Watts, eds. 2015. *Subterranean estates: Lifeworlds of oil and gas.* Ithaca, NY: Cornell University Press.

Apter, Andrew. 2005. *Pan African nation: Oil and the spectacle of culture in Nigeria.* Chicago: Chicago University Press.

Ballestero, Andrea. 2015. The ethics of a formula: Calculating a financial-humanitarian price for water. *American Ethnologist* 42(2): 262–278.

Barak, On. 2009. Scraping the surface: The techno-politics of modern streets in turn-of-twentieth-century Alexandria. *Mediterranean Historical Review* 24: 187–205.

———. 2013. *On time: Technology and temporality in modern Egypt.* Berkeley, CA: University of California Press.

Barnes, Jessica. 2014. *Cultivating the Nile: The everyday politics of water in Egypt.* Durham: Duke University Press.

Barry, Andrew. 2001. *Political machines.* London: Athlone.

———. 2013. Material politics: Disputes along the pipeline. Chichester, UK: Wiley-Blackwell.

Bear, Laura. 2015. *Navigating austerity: Currents of debt along a South Asian river.* Stanford, CA: Stanford University Press.

Bear, Laura, Karen Ho, Anna Tsing, and Sylvia Yanagisako. 2015. "Gens: A feminist manifesto for the study of capitalism." Theorizing the contemporary, *Cultural Anthropology* website, March 30, 2015. https://culanth.org/fieldsights/652-gens-a-feminist-manifesto-for-the-study-of-capitalism.

Behrent, M. C. 2013. Foucault and technology. *History and Technology* 29: 54–104.

Benjamin, Solly. 2005. Touts, pirates and ghosts. In *Bare acts*, Sarai Reader 05, 242–254. Williamsberg, NY: Autonomedia.

Bennett, Jane. 2010. *Vibrant matter: A political ecology of things.* Durham: Duke University Press.

Bowker, Geoffrey C. 1994. *Science on the run: Information management and industrial geophysics at Schlumberger, 1920–1940.* Cambridge, MA: MIT Press.

———. 2015. Temporality. Theorizing the contemporary, *Cultural Anthropology* website, September 24, 2015. https://culanth.org/fieldsights/723-temporality.

Boyer, Dominic. 2014. Energopower: An introduction. *Anthropological Quarterly* 87: 309–333.

Braun, Bruce. 2005. Environmental issues: Writing a more-than-human urban geography. *Progress in Human Geography* 29: 635–650.

Braun, Bruce, and Sarah Whatmore. 2011. *Political matter: Technoscience, democracy, and public life.* Minneapolis: University of Minnesota Press.

Bridge, Gavin and Philippe Le Billon. 2013. *Oil.* Cambridge: Polity Press.

Caldeira, Teresa Pires do Rio. 2000. *City of walls: Crime, segregation, and citizenship in São Paulo.* Berkeley, CA: University of California Press.

Callon, Michel, 1984. Some elements of a sociology of translation: Domestication of the scallops and the fishermen of St Brieuc Bay. *The Sociological Review*, 32(1_suppl): 196–233.

Campbell, Jeremy. 2012. Between the material and the figural road: The incompleteness of colonial geographies in Amazonia. *Mobilities* 7: 481–500.

Carse, Ashley. 2014. *Beyond the big ditch: Politics, ecology, and infrastructure at the Panama Canal.* Cambridge, MA: MIT Press.

Certeau, Michel de. 2002. *The practice of everyday life.* Berkeley, CA: University of California Press.

Chakrabarty, Dipesh. 2000. *Provincializing Europe: Postcolonial thought and historical difference.* Princeton, NJ: Princeton University Press.

———. 2009. The climate of history: Four theses. *Critical Inquiry*, 35(2), 197–222.

Chatterjee, Partha. 2001. On civil and political society in post-colonial democracies. In *Civil society: History and possibilities*, ed. Sudipta. Kaviraj and Sunil. Khilnani, 165–178. New York: Cambridge University Press.

———. 2004. *The politics of the governed: Reflections on popular politics in most of the world.* New York: Columbia University Press.

Chu, Julie Y. 2014. When infrastructures attack: The workings of disrepair in China. *American Ethnologist* 41(2): 351–367.

Collier, Stephen. 2011. *Post-Soviet social: Neoliberalism, social modernity, biopolitics.* Princeton, NJ: Princeton University Press.

Coronil, Fernando. 1997. *The magical state: Nature, money, modernity in Venezuela*. Chicago: Chicago University Press.

Coutard, Olivier, Richard Hanley, and Rae Zimmerman, eds. 2004. *Sustaining urban networks: The social diffusion of large technical systems*. New York: Routledge.

Cramer, Jon and Rada Katsarova. 2015. Race, class, and social reproduction in the urban present: The case of the Detroit water and sewage system. *Viewpoint Magazine* (5). https://www.viewpointmag.com/2015/10/31/race-class-and-social-reproduction-in -the-urban-present-the-case-of-the-detroit-water-and-sewage-system.

De Boeck, Filip. 2011. Inhabiting ocular ground: Kinshasa's future in the light of Congo's spectral urban politics. *Cultural Anthropology* 26(2): 263–286.

———. 2015. "Divining" the city: Rhythm, amalgamation and knotting as forms of urbanity. *Social Dynamics: A Journal of African Studies* 41(1).

de la Cadena, Marisol. 2010. Indigenous cosmopolitics in the Andes: Conceptual reflections beyond "politics". *Cultural Anthropology* 25(2): 334–370.

Degani, Michael. 2013. Emergency power: Time, ethics and electricity in postsocialist Tanzania. In *Cultures of energy: Power, practices, technologies*, ed. S. Strauss, S. Rupp, and T. F. Love, 177–193. Walnut Creek, CA: Left Coast Press.

Edwards, Paul. 2003. Infrastructure and modernity: Force, time and social organization in the history of sociotechnical systems. In *Modernity and technology*, ed. T. Misa, P. Brey, and A. Feenberg, 185–225. Cambridge, MA: MIT Press.

Edwards, Paul, Geoffrey Bowker, Steven Jackson, and Robin Williams. 2009. Introduction: An agenda for infrastructure studies. *Journal of the Association for Information Systems* 10(5): 365–374.

Elster, Jon. 1983. *Explaining technical change: A case study in the philosophy of science*. Cambridge: Cambridge University Press.

Elyachar, Julia. 2010. Phatic labor, infrastructure, and the question of empowerment in Cairo. *American Ethnologist* 37: 452–464.

Engels, Frederick. (1884) 2010. The origin of the family, private property, and the state. Marxists Internet Archive. https://www.marxists.org/archive/marx/works/download /pdf/origin_family.pdf.

Escobar, Arturo. 1995. *Encountering development: The making and unmaking of the third world*. Princeton, NJ: Princeton University Press.

Fanon, Frantz. 1961. *The Wretched of the Earth*. New York: Grove.

Fassin, Didier. 2011. Coming back to life: An anthropological reassessment of biopolitics and governmentality. In *Governmentality: current issues and future challenges*, ed. U. Bröckling, S. Krasmann, and T. Lemke, 185–200. New York: Routledge.

Fennell, Catherine. n.d. Wasted house, leaded world. Unpublished manuscript, last modified November 11, 2016. Microsoft word file.

Ferguson, James. 1994. *The anti-politics machine: "Development," depoliticization, and bureaucratic power in Lesotho*. Minneapolis: University of Minnesota Press.

———. 1999. *Expectations of modernity: Myths and meanings of urban life on the Zambian Copperbelt*. Berkeley, CA: University of California Press.

———. 2012. Structures of responsibility. *Ethnography* 13: 558–562.

————. 2015. *Give a man a fish: Reflections on the new politics of distribution*. Durham: Duke University Press.

Furlong, Kathryn. 2014. STS beyond the "modern infrastructure ideal": Extending theory by engaging with infrastructure challenges in the South. *Technology in Society* 38: 139–147.

Geertz, Clifford. 1972. The wet and the dry: Traditional irrigation in Bali and Morocco. *Human Ecology* 1(1): 23–37.

Gidwani, Vinay. 2008. *Capital, interrupted: Agrarian development and the politics of work in India*. Minneapolis: University of Minnesota Press.

Gilmore, Ruth Wilson. 2007. *Golden gulag: Prisons, surplus, crisis, and opposition in globalizing California*. Berkeley, CA: University of California Press.

Gordillo, Gastón. 2014. *Rubble: The afterlife of destruction*. Durham: Duke University Press.

Graham, Stephen. 2002. Bulldozers and bombs: The latest Palestinian-Israeli conflict as asymmetric urbicide. *Antipode* 34: 642–649. https://doi.org/10.1111/1467-8330.00259.

————. 2010. When infrastructures fail. In *Disrupted cities: When infrastructure fails*, ed. Stephen Graham, 1–26. New York: Routledge.

Graham, Stephen, and Simon Marvin. 2001. *Splintering urbanism: Networked infrastructures, technological mobilities and the urban condition*. New York: Routledge.

Graham, Stephen, and Colin McFarlane. 2015. *Infrastructural lives: Urban infrastructure in context*. New York: Routledge.

Graham, Stephen, and Nigel Thrift. 2007. Out of order: Understanding repair and maintenance. *Theory, Culture and Society* 24: 1–25.

Gupta, Akhil. 1992. The reincarnation of souls and the rebirth of commodities: Representations of time in "East" and "West". *Cultural Critique* 22: 187–211.

————. 1998. *Postcolonial developments: Agriculture in the making of modern India*. Durham: Duke University Press.

————. 2012. *Red tape: Bureaucracy, structural violence, and poverty in India*. Durham: Duke University Press.

Gupta, Akhil, and James Ferguson. 1992. Beyond "culture": Space, identity, and the politics of difference. *Cultural Anthropology* 7(1): 6–23.

Habermas, Jürgen. 1989. *The structural transformation of the public sphere: An inquiry into a category of bourgeois society*. Cambridge, MA: MIT Press.

Haraway, Donna J. 1991. *Simians, cyborgs, and women: The reinvention of nature*. New York: Routledge.

————. 2003. *The companion species manifesto: Dogs, people, and significant otherness*. Chicago: Prickly Paradigm.

————. 2016. *Staying with the trouble: Making kin in the Chthulucene*. Durham: Duke University Press.

Harris, Marvin. 1966. The cultural ecology of India's sacred cattle. *Current Anthropology* 7(1): 51–56.

Harrison, Virginia. 2013. World's best economies in 2013. *CNN Online*. http://money.cnn.com/gallery/news/economy/2013/12/27/best-economies/2.html.

Harvey, David. 1989. *The condition of postmodernity: An enquiry into the origins of cultural change*. Oxford: Blackwell.

———. 2005. *A brief history of neoliberalism*. New York: Oxford University Press.

Harvey, Penelope. 2010. Cementing relations: The materiality of roads and public spaces pn provincial Peru. *Social Analysis* 54: 28–46.

———. 2015. Materials. Theorizing the contemporary, *Cultural Anthropology* website, September 24, 2015. https://culanth.org/fieldsights/719-materials.

Harvey, Penelope, and Hannah Knox. 2015. *Roads: An anthropology of infrastructure and expertise*. Ithaca, NY: Cornell University Press.

Hetherington, Kregg. 2014. Waiting for the surveyor: Development promises and the temporality of infrastructure. *Journal of Latin American and Caribbean Anthropology* 19: 195–211.

Highsmith, Andrew R. 2015. *Demolition means progress: Flint, Michigan and the fate of the American metropolis*. Chicago: University of Chicago Press.

Hughes, Thomas Parke. 1983. *Networks of power: Electrification in Western society, 1880–1930*. Baltimore, MD: Johns Hopkins University Press.

Humphrey, Caroline. 2005. Ideology in infrastructure: Architecture and Soviet imagination. *Journal of the Royal Anthropological Institute* 11: 39–58.

Inda, Jonathan Xavier. 2005. *Anthropologies of modernity: Foucault, governmentality, and life politics*. Malden, MA: Blackwell.

Ingold, Tim. 2012. Toward an ecology of materials. *Annual Review of Anthropology* 41: 427–442.

Jackson, Steven. 2014. Rethinking repair. In *Media technologies: Essays on communication, materiality and society*, ed. T. Gillespie, P. Boczkowski, and K. Foot, 221–239. Cambridge, MA: MIT Press.

Jensen, Casper Bruun. 2017. Pipe dreams: Sewage infrastructure and activity trails in Phnom Penh. *Ethnos* 82(4): 1–21.

———. n.d. Mekong scales: Domains, test-sites and the micro-uncommons. Unpublished paper. https://www.academia.edu/11623322/Mekong_Scales_Domains_Test-Sites_and _the_Micro-Uncommons.

Joyce, Patrick. 2003. *The rule of freedom: Liberalism and the modern city*. London: Verso.

Kale, Sunila S. 2014. *Electrifying India: Regional political economies of development*. Stanford, CA: Stanford University Press.

Kaviraj, Sudipta. 2001. In search of civil society. In *In search of civil society*, ed. Sudipta Kaviraj and Sunil Khilnani, 287–323. New York: Cambridge University Press.

Kohn, Eduardo. 2013. *How forests think: Toward an anthropology beyond the human*. Berkeley, CA: University of California Press.

Kothari, Ashish, and Rajiv Bhartari. 1984. Narmada Valley project: Development or destruction. *Economic and Political Weekly* 19(23): 907–920.

Lansing, John Stephen. 1991. *Priests and programmers: Technologies of power in the engineered landscape of Bali*. Princeton, NJ: Princeton University Press.

Larkin, Brian. 2008. *Signal and noise: Media, infrastructure, and urban culture in Nigeria*. Durham: Duke University Press.

———. 2013. The politics and poetics of infrastructure. *Annual Review of Anthropology* 42:327–343.

———. 2015. Form. Theorizing the contemporary, *Cultural Anthropology* website, September 24, 2015. https://culanth.org/fieldsights/718-form.

Latour, Bruno. 1996. *Aramis, or the love of technology*. Cambridge, MA: Harvard University Press.

———. 2012. *We have never been modern*. Cambridge, MA: Harvard University Press.

Law, John. 1987. *The social construction of technological systems: New directions in the sociology and history of technology*, ed. W. Bijker, T. P. Hughes, and T. Pinch. Cambridge, MA: MIT Press.

Lefebvre, Henri. 1991. *The production of space*. Cambridge, MA: Blackwell.

Lemke, Thomas. 2014. New materialisms: Foucault and the "government of things." *Theory, Culture and Society.* 32(4): 3–25.

Limbert, Mandana E. 2010. *In the time of oil : Piety, memory, and social life in an Omani town*. Stanford, CA: Stanford University Press.

Marcus, George E., and Michael M. J. Fischer. 1986. *Anthropology as cultural critique: An experimental moment in the human sciences*. Chicago: University of Chicago Press.

Mauss, Marcel. 2008. "Seasonal Variations of the Eskimo". In *Environmental anthropology: A historical reader*, ed. Michael R. Dove and Carol Carpenter, 157–167. Malden, MA: Blackwell.

Marres, Noortje. 2012. *Material participation: Technology, the environment and everyday publics*. Houndmills, UK: Palgrave Macmillan.

Maurer, Bill. 2012. Payment: Forms and functions of value transfer in contemporary society. *Cambridge Anthropology* 30(2): 15–35.

McFarlane, C., and J. Rutherford. 2008. Political infrastructures: Governing and experiencing the fabric of the city. *International Journal of Urban and Regional Research* 32: 363–374.

McKittrick, Katherine. 2011. On plantations, prisons, and a black sense of place. *Social & Cultural Geography* 12(8): 947–963.

Meehan, Katie. 2014. Tool-power: Water infrastructure as wellsprings of state power. *Geoforum* 57: 215–224.

Mehta, Lyla. 2005. *The politics and poetics of water: Naturalising scarcity in western India*. New Delhi: Orient Longman.

Melamed, Jodi. 2006. The spirit of neoliberalism: From racial liberalism to neoliberal multiculturalism. *Social Text* 24(4(89)): 1–24. https://doi.org/10.1215/01642472-2006-009.

Mills, Charles Wright. 2011. Liberalism and the racial state. In *State of white supremacy: Racism, governance, and the United States*, ed. Moon-Kie Jung, João H. Costa Vargas, and Eduardo Bonilla-Silva. Stanford, CA: Stanford University Press.

Mitchell, Timothy. 2002. *Rule of experts: Egypt, techno-politics, modernity*. Berkeley, CA: University of California Press.

———. 2011. *Carbon democracy: Political power in the age of oil*. London: Verso.

Mol, Annemarie. 2002. *The body multiple : Ontology in medical practice, science and cultural theory*. Durham: Duke University Press.

Morgan, Lewis Henry. (1877) 2004. *Ancient society, or researches in the lines of human progress from savagery through barbarism to civilization*. Marxists Internet Archive. http://www.marxists.org/reference/archive/morgan-lewis/ancient-society.

Mrázek, Rudolf. 2002. *Engineers of happy land: Technology and nationalism in a colony*. Princeton, NJ: Princeton University Press.

Nugent, David. 2004. Governing states. In *A companion to the anthropology of politics*, ed. David Nugent and Joan Vincent, 198–215. Malden, MA: Blackwell.

Ranganathan, Malini. 2016. Thinking with Flint: Racial liberalism and the roots of an American water tragedy. *Capitalism Nature Socialism* 27(3): 17–33. https://doi.org/10 .1080/10455752.2016.1206583.

Redfield, Peter. 2005. Foucault in the tropics: Displacing the panopticon. In *Anthropologies of modernity: Foucault, governmentality, and life politics*, ed. Jonathan Xavier Inda, 50–79. Malden, MA: Blackwell.

Ribeiro, Gustavo Lins. 1994. *Transnational capitalism and hydropolitics in Argentina: The Yacyretá high dam*. Gainesville: University Press of Florida.

Rodgers, Dennis, and Bruce O'Neill. 2012. Infrastructural violence: Introduction to the special issue. *Ethnography* 13: 401–412.

Rosenberg, Nathan. 1976. *Perspectives on technology*. Cambridge: Cambridge University Press.

Rostow, W. W. 1960. *The stages of economic growth, a non-Communist manifesto*. Cambridge: Cambridge University Press.

Sassen, Saskia. 1991. *The global city: New York, London, Tokyo*. Princeton, NJ: Princeton University Press.

Schwenkel, Christina. 2013. Post/socialist affect: Ruination and reconstruction of the nation in urban Vietnam. *Cultural Anthropology* 28: 252–277.

———. 2015a. Spectacular infrastructure and its breakdown in socialist Vietnam. *American Ethnologist* 42(3): 520–534.

———. 2015b. Sense. Theorizing the contemporary, *Cultural Anthropology* website, September 24, 2015. https://culanth.org/fieldsights/721-sense.

Scott, James. 1998. *Seeing like a state: How certain schemes to improve the human condition have failed*. New Haven, CT: Yale University Press.

Sharma, Aradhana. 2008. *Logics of empowerment : Development, gender, and governance in neoliberal India*. Minneapolis: University of Minnesota Press.

Sheth, Falguni. 2009. Toward a political philosophy of race. New York: SUNY Press.

Simone, AbdouMaliq. 2004. People as infrastructure: Intersecting fragments in Johannesburg. *Public Culture* 16(3): 407–429.

———. 2012. Infrastructure: Introductory commentary. Curated collections, *Cultural Anthropology* website, November 26, 2012. https://culanth.org/curated_collections /11-infrastructure/discussions/12-infrastructure-introductory-commentary-by-abdou maliq-simone.

Star, Susan Leigh. 1999. The ethnography of infrastructure. *American Behavioral Scientist* 43: 377–391.

Star, Susan Leigh, and Karen Ruhleder. 1996. Steps toward an ecology of infrastructure: Design and access for large information spaces. *Information systems research* 7(1): 111–134.

Starosielski, Nicole. 2015. *The undersea network*. Durham: Duke University Press.

Steward, Julian. 1955. *Theory of culture change: The methodology of multilinear evolution.* Urbana: University of Illinois Press.

Stoler, Ann Laura. 2013. *Imperial debris: On ruins and ruination*. Durham: Duke University Press.

Sundaram, Ravi. 2010. *Pirate modernity: Delhi's media urbanism*. New York: Routledge.

Thrift, Nigel. 2005. *Knowing capitalism*. London: Sage.

Tsing, Anna. 2005. *Friction: An ethnography of global connection*. Princeton, NJ: Princeton University Press.

von Schnitzler, Antina. 2013. Traveling technologies: Infrastructure, ethical regimes, and the materiality of politics in South Africa. *Cultural Anthropology* 28: 670–693.

———. 2015. Ends. Theorizing the contemporary, *Cultural Anthropology* website, September 24, 2015. https://culanth.org/fieldsights/713-ends.

———. 2016. *Democracy's infrastructure: Techno-politics and protest after Apartheid.* Princeton University Press.

Warner, Michael. 2002. Publics and counterpublics. *Public Culture* 14: 49–90.

Weizman, Eyal. 2012. *Hollow land: Israel's architecure of occupation*. New York: Verso.

Weszkalnys, Gisa. 2014. Anticipating oil: The temporal politics of a disaster yet to come. *Sociological Review* 62(S1): 211–235.

White, Leslie. 1943. Energy and the Evolution of a Culture. *American Anthropologist*, 45(3): 335–356.

Williams, Raymond. 1973. *The country and the city*. New York: Oxford University Press.

Winner, Langdon. 1999. Do artifacts have politics? In *The social shaping of technology*, ed. D. A. MacKenzie and J. Wajcman, 26–39. Buckingham, UK: Open University Press.

Wolf-Meyer, Matthew. 2012. *The slumbering masses: Sleep, medicine, and modern American life*. Minneapolis: University of Minnesota Press.

PART I. *Time*

1

Infrastructural Time

HANNAH APPEL

Malabo es la metáfora más limpia de los desafíos y oportunidades de Guinea Ecuatorial. Pero Malabo no es una magdalena, en Malabo hoy no se busca el tiempo perdido, se construye un nuevo tiempo, un tiempo que será o no será rotundo, un tiempo que pueda que esté sujeto aquí y allá y pueda también que no, un tiempo que basculará entre un ayer inventado o aclarado y una mañana deseado o merecido. En estos momentos, ahora mismo, Malabo, la ciudad remordida, está viviendo un sueño de progresos y retrocesos, de redenciones y corrupciones, un sueño de buenas esperanzas, un sueño de edificios de vidrio y azulejos, un sueño de rotondas para dar la vuelta al mundo en ochenta días y la vuelta al día en ochenta mundos, un sueño de lenguas de alquitrán que hacen vibrar a árboles milenarios, un sueño de móviles para todos, agua para todos, electricidad para todos, un sueño que todos quieren soñar para que sea un sueño para todos y por todos, para que no haya desalojos sin esperanzas, para que el estado no sea Goliat pues no todos los ciudadanos pueden ser David . . .

Malabo is the clearest metaphor of Equatorial Guinea's challenges and opportunities. But Malabo is not a madeleine. Today Malabo is not in search of things past, but building a new time, a time that may or may not be decisive, a time that might be held here and there, or that might not, a time that will oscillate between a made-up or made-clear yesterday and a wished-for or well-deserved tomorrow. At this moment, right now, Malabo, the regretful city, is living a dream of progresses and relapses, of redemptions and corruptions, a dream of good hopes, a dream of glass and tile buildings, a dream of roundabouts to go around the world in eighty days and around the day in eighty worlds, a dream of tar tongues that make ancient trees vibrate, a dream of mobile phones for everybody, water for everybody, electricity for everybody, a dream that everybody wants to dream so that it might become a dream to everybody and for everybody, so that there won't be evictions without hope, so that the state won't be Goliath, for not all citizens can be David . . .

—CÉSAR A. MBA ABOGO (2011)

The first months of 2008 were dark in Malabo, Equatorial Guinea's capital. The city went for days on end without electricity, stretching at one point to two weeks. Those who could afford it used private generators in the days *sin luz* (literally, without light) to keep businesses running, to keep food cool, or to allow electric light, recorded music, or television. The city filled with the clattering roar of generator motors fighting their flimsy steel containers and the stench of diesel exhaust. My neighbors—a Lebanese-owned restaurant and nightclub complex—had a powerful generator, the noise and fumes from which, on occasion, filled my small apartment so completely that staying inside became unbearable. Unable to sleep on one such generator-filled night, I opened my door to look for air and to share water and complaints with Moussa, the Senegalese watchman who spent every night on the sidewalk outside the Lebanese complex. We chatted about the blackout. He said that Senegal provides electricity for many of its neighbors—for Guinea Bissau, for as far away as the Ivory Coast. We laughed and said that Senegal should consider providing electricity to Equatorial Guinea as well. But for all Senegal's apparent success in the realm of electricity provision, Moussa spent every night sleeping on cardboard laid over broken concrete on Malabo's sidewalk, inhaling generator fumes, covered head to toe in clothing and plastic sheeting to fend off malarial mosquitoes in the eighty-plus-degree heat. Even without electricity, sleeping on the sidewalk in Equatorial Guinea, he seemed to think, provided better prospects than his native Senegal.

The oil and gas business had come to Equatorial Guinea roughly ten years before Moussa's 2007 arrival, and with it came a series of ambiguous distinctions: Equatorial Guinea has been the world's fastest growing economy; it is now the wealthiest country per capita in Africa; and in 2013 Equatorial Guinea saw more investment as a percentage of GDP than any other country in the world, at 61.3 percent (Harrison 2013). It is eminently reasonable to assume, as Moussa did, that even sleeping on the sidewalk where the streets are paved with gold might get you a little closer to it.

"Investment as a percentage of GDP" is the statistical reflection of infrastructure projects in the national economy form. The statistic accounts for any investment in construction of roads, railways, electric and water grids, schools, hospitals, commercial and industrial buildings, and beyond. That Equatorial Guinea had the highest investment percentage in the world in 2013 reflects the extraordinary intensity of infrastructure development there. In a country the size of Delaware with roughly 750,000 inhabitants, new infrastructure saturated daily life. For Malabo's residents, the experience of these projects was visceral, sensory (Mrázek 2002; Larkin 2013)—the endless thrum of jackhammers, bulldozers, and trucks too big for old colonial roads; the air full of cement dust that settles on skin and in mouths. Close your eyes, and there is a new skyscraper when you open them. Construction projects set up haphazardly in the middle of everything buzzed with day laborers—often new immigrants like Moussa from Senegal, Cameroon, or Benin—welding, swinging metal beams, digging ditches that drop into the bowels of the old colonial undercity. Central Malabo is small enough that all pedestrians must walk directly through these sites on their way here or there, hoping not to get sprayed by welding spatter or fall into a ditch.

The infrastructure frenzy stretched unevenly outside Malabo into Luba, Riaba, and Moka; into the continental region in Bata, Mongomo, and Oyala; and even to the long-neglected island of Annobón, where the Moroccan company SOMAGEC has built a new airport, hotel, and road system. Chinese construction workers lingered, smoking on the edges of worksites across the country. Thousands of workers from China Dalian, China Communications Construction Company, and others paved roads and built bridges and dams. Arab Contractors—Egypt's largest parastatal construction firm—also had workers throughout Equatorial Guinea, building a stadium here, a governmental palace or ministry there.

The infrastructure boom meant a landscape not only covered by cranes and scaffolding, etched with ditches and quarries, but also swathed with elaborate signs of futurity—colorful depictions of the infrastructural project that would soon stand in this or that patch of newly cleared forest: a new refinery or storage plant, a water-treatment facility, a hydroelectric dam, the new BEAC Regional Bank, the new headquarters of the national oil or gas company, a series of mansions and apartment buildings (figure 1.1). Land was cleared as fast as you could blink, and signs declaring government ownership blossomed in the newly exposed red earth. Expropriation was rampant and all but incontestable except for those most intimately connected with the regime, and even for them the process was protracted, cumbersome, and most likely futile. In an almost farcical move, the president went to great lengths on television to explain how *even he* had been expropriated by this unstoppable future, as thousands of acres of his private land became "state property."[1]

Among the projects springing up in these expropriated spaces was a gridded community of small white homes with blue and red roofs stretching as far as the eye could see, said on the one hand to be subsidized by the state, intended for those who had been dispossessed, but widely known to be for sale by members of government to the highest bidder (figure 1.2). The development was named La Buena Esperanza (good hope). Indeed, when Mba writes in the epigraph

赤道几内亚巴塔成品油库
TANQUE DE ALMACENAMIENTO DE CARBURANTE EN BATA DE GUINEA ECUAT(

承建单位 ENTIDAD PARA ENCARGARSE DE LA CONSTRUCCION

FIGURE 1.1 A sign depicting future fuel-storage tank construction in Bata. Photo by the author.

that Malabo is living "a dream of good hopes," or a dream "that there won't be evictions without hope," he refers directly to the controversy of the La Buena Esperanza development.

In this article, I make explicit Mba's literary assertion that Equatorial Guinea's infrastructure boom is the construction of the memory of petroleum; that these projects are building *time and temporalities* at the same time as they are building "material forms that allow for the possibility of exchange over space," as Brian Larkin (2013: 327) has defined infrastructure. Using ethnographic material from the built environment and from Equatorial Guinea's "second" national economic conference, the article dwells on questions of developmental time—linearity, progress, teleology—on the one hand, and oil time on the other: repetition and cyclicality; serial frontiers; abandonment, decommission, and ruins. Before getting to the ethnographic material, however, a brief consideration of infrastructural time as it appears in the epigraph.

The parenthetical in Mba's title, translated as "(The construction of) the memory of petroleum," is a play on Derridean genealogies to assert a materialist insistence—that the memory of petroleum is quite literally being constructed now in the glass-skinned buildings of Malabo Dos (Appel 2012b), in

FIGURE 1.2 La Buena Esperanza housing development. Photo by the author.

the newly paved roads, in the remarkable government buildings, all of which will outlast, in one way or another, oil itself. But insofar as they were funded by oil, they will be its monuments and, arguably, its ruins. In Mba's writing, infrastructural time folds over on itself; it oscillates and stutters as progress and relapse coincide. This stuttering time reflects the fact that infrastructure does not so much "arrive" in Malabo as advance and retreat (Carse 2014): new roads are torn up for new pipes; new pipes fail to carry the water for which they were designed; the treatment plant promised on the billboard sits as a half-built ruin (Gupta, this volume; Roitman and Mbembe 1995). In September 2007, when I arrived for a twelve-month fieldwork trip, the roads that had been laid during a previous trip in the summer of 2006 were ripped up throughout the city, *because there will be running potable water by November*, the official narrative went. There was not. And still in 2017 running water in Malabo and beyond is sporadic at best and not potable. This is a particular kind of infrastructural futurity that is more akin to deferral, more akin to the way Ann Stoler (2008) has described imperial formations as "states of deferral that mete out promissory notes that are not exceptions to [the operation of such formations] but constitutive of them" (193). Regularly unfinished, and often faulty, new construction is haunted by abandonment. Infrastructure's promise of distribution

Infrastructural Time 45

is refracted through the forms of rule and inequality that precede it and create the ground on which it is built. "The architectural text of unfinished edifices stands as a reminder of . . . political subtext" (Roitman and Mbembe 1995: 335). For Mba, infrastructure becomes the stuff of dreams in this sporadic, labyrinthine temporality—roundabouts become time machines, roads are the tongues of a new creature lapping at the much older forest. A shared infrastructural modernity itself becomes a dream—"a dream of mobile phones for everybody, water for everybody, electricity for everybody." In Equatorial Guinea's infrastructure boom, futurity and deferral, teleology and cyclicality coexist in a dizzy stasis.

Specious ideas about infrastructure were always central to staged theories of modernization, from Lewis Henry Morgan's (1877) account of bows and arrows to irrigation to iron construction, to the techno-developmentalist teleologies that animated both Marxist (Engels [1884] 2010) and orthodox modernization theories (Rostow 1960). Though the staged theories in which infrastructure was mobilized have now been rejected as valid social scientific description, anthropologists have often noted that modernization theory still hangs out ethnographically (Ferguson 1999). People around the world talk in terms of developmental time, progress and relapse, of being behind and needing to catch up. In Equatorial Guinea, the modernization narrative persisted in part through the materiality of infrastructure; through the affective distinction (Mrázek 2002; Larkin 2013) between a dirt road and a paved one; between boiling water from a river or well before drinking it or drinking it directly from a tap. Where infrastructure served as a material metonym of modernity, attention to infrastructure's actual life courses confounds developmentalist narratives of linear progress. For example, on the eve of Equatorial Guinea's Independence Day—October 12, 2008—I sat around the dinner table, in darkness, with an Equatoguinean family who hosted me for several months in the field. With the electricity out and candles flickering, military planes flew low and loud overhead and our dinner conversation was cut off by the noise several times. Both the mother and the eldest brother (himself well into his forties) remarked at the irony of "advanced" military technology in a country where electricity was sporadic at best. When I excused myself from the table and wished the family "feliz fiestas" (happy holidays), the brother responded, "We're not independent yet."

In both Mba's epigraph and the dinner anecdote, the experience of infrastructure becomes a Benjaminian constellation: "It is not that what is past casts its light on what is present, or what is present its light on what is past; rather, . . . what has been comes together in a flash with the now to form a constellation . . .

dialectics at a standstill" (2002: 262). For Benjamin, this was the analytical richness of the Arcades project, the ways in which perceptions of infrastructure can "replace the linear narratives of history with a constellation of events frozen momentarily in an image containing histories of the past and present" (Kennel et al. 2009). Fighter jets with no electricity, roundabouts that go around the day in eighty worlds gesture to failures of linear time, to the affective inhabitance of a time that *feels* fractured, that seems both to explode with futurity and collapse under the weight of the postcolonial now—"a made-up or made-clear yesterday and a wished-for or well-deserved tomorrow." But rather than dwell comfortably in this folded time, both Mba and my dinner companions *insist*, through infrastructure, on what we might call developmental time, a normative temporality in which military planes *should* coincide with public electricity provision, in which the frantic construction in Malabo and beyond *should* coincide with water for everybody, electricity for everybody. Lived hybridity, sure; coeval time, sure; but, simultaneously, a materialist insistence on a desired standard of living (Ferguson 2006) that Equatoguineans routinely, and indeed metonymically, articulated through infrastructure.

Infrastructural Time I: Development and the National Economy

"Pisos son el desarrollo" (Floors are development) went the joke. Development, seemingly, can be calculated according to how many levels a building has. The joke was a not-so-subtle critique of the Equatoguinean state's apparent development vision—build, build, build. Indeed, in the contracts obligating foreign oil and gas companies to build local headquarters in Malabo Dos, one clause specified that each building had to be at least seven stories high, irrespective of how many employees the given company had in-country (Appel 2012b). Equatoguinean friends often satirized the nation's motto—"unidad, paz y justicia," or unity, peace, and justice—by changing it to "unidad, hormigón, y cristal" (unity, concrete, and glass).

Beyond jokes that circulated at bars or around dinner tables, the relationship of infrastructure to development had a lively official life, as state actors often had to justify the plainly exorbitant investment in some kinds of infrastructure above others and in infrastructure above all else. When pressed, ministers and other state representatives routinely attributed the persistent unreliability of electricity, running water, and school or healthcare facilities to the need to focus first on economic development through *other* infrastructure projects—roads, dams, ports, airports, government buildings. These projects were contracted and funded by the government, framed as the necessary first step for private sector–led growth. In other words, officials justified roads before water, airports before electricity, as a strategy to stimulate the economic growth that would, in a deferred future, enable public infrastructural provisioning. Yet "the economy" here serves as a complicated pretext for a suite of practices where the Equatoguinean state and U.S. oil investment meet (Appel 2017). First, as the International Monetary Fund (IMF) puts it, "Public investment is a direct substitute for private investment ... public investment in Equatorial Guinea includes housing, roads, ports, and airports as well as capital or subsidies for fishing, farming, and transportation" (2009: 10). In other words, reflecting the fact that Equatorial Guinea is one of only a handful of nations, including China, able to pay outright for infrastructure projects, infrastructure becomes a method to launder petrodollars. No-bid contracts attached to specific members of the regime funnel state revenue from oil and gas to foreign firms. This process often includes inflated contracts with parastatals from sympathetic regimes (China, Egypt, Morocco), the excess from which is "kicked back" to Equatoguinean officials (referred to in Equatorial Guinea as indemnizaciones: see Appel 2012b). In Cameroon's petro boom, Janet Roitman and Achille Mbembe (1995) refer to similar practices as "an extravagant and unproductive economy of public and private expenditures" giving rise to "an entire social commerce with forms of political exchange and modes of appropriating public goods that were widely known" (334–335). Perhaps needless to say, nothing easily recognizable as a "private sector" emerges from this form of infrastructure-focused appropriation, though economic growth remains the official explanation for projects including the €50 million Bata-Mbini Bridge or the 148,000-square-foot Sipopo Conference Center, completed by the Turkish firm Tabanlıoğlu Architects, cost withheld. Infrastructure as text; economy as pretext; politics as subtext.

Even as pretext, however, in Equatorial Guinea as elsewhere the economy in all its *agencement* (its conceptual and material work in the world; the ways in which that work is imagined and described; the other objects and concepts to which it is connected) is a privileged object, perhaps *the* privileged object (Appel 2017). The future is imagined around it; normative policy thinking takes it as its purpose, as do development programs and local politicians. In 1983, after his third trip to Equatorial Guinea, UN special rapporteur and Costa Rican law professor Fernando Volio Jimenez, reported that "one official after another all the way up to Obiang [Obiang Nguema Mbasogo, the president] himself" justified the limitations on the press (there was none) and on political participation (political parties were banned) "as being necessary for the focusing of attention on economic issues" (Fegley 1991: 220). Over twenty years later during my time in Equatorial Guinea, ministers and other state representatives routinely attributed the tenacious problems of public infrastructure, not to mention the continued draconian limits on press and political organizing of any kind, to the need to focus first on infrastructure. The economy and the infrastructure that would "create" it were both objects of the future and justifications of the present's constant deferral. Here it is not the nation but the national economy—its growth, diversification, and competitive advantage—that forms the object of development and developmental time. Infrastructure is the imagined materialization of this thing called an economy.

The official articulation of this vision took place at a national economic conference held in 2007, at which the state introduced their national development plan entitled Horizonte 2020 (Horizon 2020). Four months into my fieldwork and exactly ten years after what state organizers had called the country's "first" national economic conference, this was the country's "second" national economic conference.[2] While I discuss this conference and the national economy as objects of ethnographic inquiry at length elsewhere (Appel 2017) what interests me here is the official articulation of developmental time and infrastructure that was one of the conference's central themes. The Horizon 2020 conference was held in Bata, Equatorial Guinea's second city. Unlike claustrophobic Malabo, where you can barely tell that the ocean is just over there unless you peer over a wall, Bata feels expansive, with wide roads, fewer people, and room to breathe. Though daily activities of governance take place in Malabo, the public administration also moves, working periodically from a second set of ministerial and administrative buildings and palaces in Bata. (Pending the completion of Oyala—"a new future capital"—the practices and geographies of governance may move again.) At the conference, the president and prime ministers, ministers of government, oil company representatives and their local government liaisons, national and international businesspeople hoping to invest, the World Bank, the IMF, United States Agency for International Development, Washington, D.C. lobbyists, United Nations Development Program personnel, European Community and Spanish Cooperation delegates, diplomats from throughout the region, local businesspeople, and who knows who else were all brought together in the name of the future, Horizon 2020. After an initial day of registration, opening ceremonies, and closed-door meetings, day two presented conference participants with four concurrent daylong sessions: infrastructure, social sector, public sector, and private sector. Anguished at having to choose between these four clearly interrelated meetings, I followed the decidedly largest portion of the crowd into the private-sector session, held in an opulent conference hall with seating for roughly five hundred people.

The long hours of the private-sector session were filled with serial presentations from the minister of mines, representatives from the state oil (GEPetrol) and gas (Sonagas) companies, and a German agro-businessman seemingly in the process of brokering a large deal with the government. Each presenter narrated a future at once utopic and surreal. "In the energy sector," the Sonagas representative intoned, "the government will build two conditioning plants, one in Bata and one in Malabo, to make gas available for local use and diminish ongoing reliance on foreign processing of our products. They will also construct a modular refinery in the country to bring down prices. By 2020, state-of-the-art gas-processing facilities in Equatorial Guinea will monetize the gas that is currently burned off in the petroleum production process not only here, but also in Cameroon and Nigeria."

> "The potential for the fishing industry is colossal," said the agro-business man, for industrial, artisanal, and farmed fish in the ocean and rivers, with an estimated local capacity of 65,000 tons per year, the equivalent of $100,000,000. . . . The government will build two industrial centers and ice factories in Malabo and Bata to service this industry. They will educate oceanographers and boat engineers. By 2020 there will be a fleet of industrial fishing ships, and an industry producing value-added products for export including salted, dried, and smoked fish, canned and packaged products, and modern fish farms. By 2020 Equatorial Guinea will be the commercial center in the region for sea products. Annobón will be the center of the industry, where women will be transformed into commercial fishers.

Gas-conditioning plants, modular refineries, processing facilities; ice factories, industrial fishing ships, export-processing zones. The visions articulated in the private-sector session were almost entirely conceptualized through potential infrastructures. The diversification of the economy away from oil meant more infrastructures of specific kinds, which in turn relied on other infrastructures. Ice, to take one example, relies on both water and electricity. These nested infrastructural concerns came up repeatedly in the presentations. *A fleet of industrial fishing ships will require a place to fix them. We'll need fabrication shops. If we're going to be a commercial center for fish, we'll need storage facilities. Cold storage facilities. What about a port? In order to be the center of the industry, Annobón would need a port.* And so infrastructure came to stand metonymically for development itself, for the private sector diversifying away from an inordinate reliance on petroleum and into a future when everyone, "even the women of Annobón," it was suggested, would be brought into a brave new infrastructural world.

What kind of developmental time is at stake here? What does it mean that there is a slippage between economic development and national development? Is the developmental time at stake for the organizers of this conference the same developmental time envisioned in earlier modernization projects, where public works and collective goals were considered the foundations of national progress? Is Horizon 2020 equivalent to the five-year plans frequently produced in the heyday of postcolonial modernization? Is the conflation of economic growth and development something new, a new kind of time? For example, are we to consider the extensive documentation that conference participants received as planning documents or speculative documents? To the extent that a plan can be gleaned from the meeting proceedings, documentation, and spotty implementation in the years that followed, that plan is, essentially, to make Equatorial Guinea a great place to invest, to attract foreign capital to replace oil when that runs out. Thus, the development plan is both speculative in itself (in that its execution is not possible given current investment conditions, on which more below) and reliant on the speculation of others, aimed to seduce speculative capital. It is not a road map or a five-year plan, but perhaps something closer to what Guyer (2007: 410) has called "fantasy futurism and enforced presentism" in which the specificity of the near future is evacuated in favor of a distant utopia. This feels somewhat akin to Stoler's deferral, the deferral of Mba's shared modernity into a fantasy future that only reinforces the inevitability of the present.

One of the first audience members to ask a question after the surreal narration of future utopias was Alberto, the head of the Equatoguinean delegation for small- and medium-size businesses, who was the owner of a small store that sold computer equipment and, coincidentally, was also my upstairs neighbor. Obliquely referencing the amnesias of the current conference, he reminded the audience of a 1982 national forum on the promotion of business, and of the 1997 conference, where it was also established that the private sector would be the motor of development. After these reminders, Alberto asked, "Where are we now? We need financing and access to credit, education, good labor conditions, access to technology, and a fiscal climate according to the law. I'm no xenophobe," he continued, "but foreign companies are granted all the big contracts, and then we're hired at dismal wages as subcontractors. I have the capacity but I lack the capital. If I had the money, I too could subcontract an architect or an engineer." When he finished, the entire conference room erupted in boisterous applause and full-throated cheers, and he turned and smiled and waved at the crowd. His question was a thinly veiled critique of the state, and the repetitive conferences that came to the same conclusions yet produced nothing. If all the presenters were going to do was to justify further infrastructure development, his question implied, and to continue to hand those lucrative contracts to foreign companies, then at least cut locals (beyond government officials) in on a piece of the deal. But Alberto's question also worked on a second register, referring to *other* infrastructures not conventionally understood as such that enable the construction of a port or airport.

Finance and credit, education, access to technology, and a fiscal climate according to law might all be considered the "recursive and dispersed" substrates of visible infrastructure (Larkin 2013: 330; Appel and Kumar 2015), required for its construction and maintenance. In other words, Alberto is not merely saying that locals need to be cut in on lucrative infrastructure deals. Rather, he is pointing out that, in addition to a long-standing colonial tradition in which foreign companies are willing participants in the patron-client networks that infrastructural contracts infuse with cash, those companies also have access to *external* infrastructures—finance and credit—that are not available to Equatoguineans. The government contracts directly with these foreign companies in part because of the large indemnizaciones that state appointees stand to receive on any given project, but also in part because these companies have access to the financial and technological infrastructures that subtend any major construction project, infrastructures that still reside in an ever-deferred Equatoguinean future, as Alberto points out. "Given the ever-proliferating networks that can be mobilized to understand infrastructures," Larkin (2013) writes, "we are reminded that discussing an infrastructure is a categorical act. It is a moment of tearing into those heterogeneous networks to define which aspect of which network is to be discussed and which parts will be ignored" (329–330). In asking his question, Alberto was tearing into the recursive nature of what can be said to "count" as infrastructure, pointing out that access to finance, credit, law, technology, and education (almost completely unavailable to local actors) is infrastructural to the building boom by which everyone was surrounded but from which few were benefiting. Alberto and the cheering crowd with him were demanding that those present account for those infrastructures as well or, more accurately, their absence.

After the futuristic confabulations of the main conference proceedings, all participants returned to the luxurious auditorium for a third day of concluding summaries. On this closing day conference participants seemed noticeably tired. People straggled into the conference hall late; there seemed to be fewer animated discussions in the hallways. Having spent the evenings between conference days discussing the historic event over dinners with Equatoguinean friends, our conversations, too, had grown tired, moving from early laughter at the impracticable goals toward the torpor of long days and wasted time. After participants slowly filled the large hall that morning, we sat and sat, waiting for the president to appear, slowly sinking down in our chairs, wrinkling our clothing, and trying to hush our growling stomachs. I wrote in my notebook, *the future is exhausting.* But, finally, the voice of the president's protocol roused all of us from our slumps: "Su Excelencia, Jefe Del Gobierno . . ." and the president's Moroccan security guards strode in before him. The audience clapped in rhythm and Obiang joined in as he walked to his seat in the center of dignitaries and government ministers at the head table.

And then, it was as if for a moment the impossibly heavy future gave way, and the present's dystopia slipped in. In front of the president, and other members of that front table, government presenters elaborated with a bluntness that surprised me the serious and obvious problems the country will have to overcome to achieve their future goals: *We are essentially without all basic social services. There is little to no running water, none of it potable. Electricity is sporadic in the cities and not distributed throughout the territory. The health sector is essentially nonexistent for the majority of the country's residents and the education sector is little better. There is no transparent access to credit for small businesses, and no regularized process according to which one might start a business. There is a total lack of legal instruments or regulation in any and all sectors. There are serious problems with private property and contract law. During this time of infrastructure development, there is no state contract law or set of laws regarding the quality of work done.*

Infrastructure, and metonymically the economy and development itself, becomes futurity and deferral at once. The glinting steel and smooth asphalt promise of the new seem to concretize, to make more intransigent, even to justify relations between state and citizen, foreign and domestic, opportunity and foreclosure that feel, to many Equatoguineans, vestigial, outdated, plainly farcical if they were not so consequential. Here, we arrive again at Stoler's states of deferral, in which infrastructures "mete out promissory notes that are not exceptions to their operation but constitutive of them: imperial guardianship, trusteeships, delayed autonomy, temporary intervention, conditional tutelage" (2008: 193). One might object to the use of Stoler here, who was in fact referring to imperial projects as states of deferral. But we need only widen the ecology of infrastructure at stake in contemporary Equatorial Guinea to respond to that concern. It is the infrastructures of the transnational oil and gas industry—its rigs and platforms, floating production, storage and offloading vessels, corporate and residential enclaves (Appel 2012a, 2012b)—that produce first the hydrocarbons and then the rents to the Equatoguinean state that make Equatorial Guinea's infrastructure boom possible. And the oil and gas industry in Equatorial Guinea is nothing if not a temporary intervention, a convulsion of exploration and abandonment shaped not by linear teleologies of developmental time, but by cyclical, stuttering temporalities of resource discovery, exploitation, and exhaustion.

Infrastructural Time II: Depletion, Decommission, Abandonment

In the Time of Oil (Limbert 2010) traces the ways in which becoming an oil-exporting state engenders an encompassing preoccupation with oil's temporal limits—constant anxiety about a radically different near-future without oil. Between pasts and futures without oil, "the present"—the time of oil— "becomes an anomalous time ... not a step along a trajectory of infinite progress, but an interlude, surprisingly and perhaps miraculously prosperous" (Limbert 2010: 9; see also Ferguson 1999).

In Equatorial Guinea, the first and second national economic conferences took place in this self-consciously "anomalous" time, an interlude of spectacular accumulation that everyone acknowledged as such—futurity and exhaustion, exploration and abandonment were always-already coupled. The teleologies of modernization circulating in Equatorial Guinea, the habitations of developmental time, quite literally contained their own horizons, delimited by the exhaustion of oil and gas resources within thirty years, according to most estimates. Much of the state's official work in the interim, then, was to build as many buildings as possible, "using modern and permanent materials" (Ministry of Mines, Industry, and Energy 2008: 26) before new kinds of time set in—of dwindling resources, fleeing foreign capital, decommission and abandonment, of ruins. Equatoguineans needed only look across an arbitrarily drawn line on a map to Gabon to envision these ruins of the future (Gupta, this volume).

The town of Mongomo sits at the eastern edge of Equatorial Guinea's continental region. On Mongomo's own eastern edge the paved road stops abruptly at a simple barrier arm gate and kiosk that mark the border with Gabon. Past the arm gate, dirt road resumes. Oyem, some forty-one kilometers into Gabon on alternately unpaved and paved roads, is a city of forty thousand people that also knew a petroleum boom. Once-grand government buildings, functioning electrical grids, even a national railway built with Gabon's petro dollars of the 1970s and 1980s now crumble quietly, not an hour away from their apparent successors in Equatorial Guinea. Gabon's petro-boom coincided with a time of mass violence in Equatorial Guinea, and many Equatoguineans fled into exile in Gabon and lived there during the country's fleeting moments of oil-fueled developmental time. In Gabon in the 1970s and 1980s, Equatoguineans described to me being treated "like dogs," the poor refugee neighbors fleeing a violent and capricious state. As its infrastructure crumbled across a mere line on a map, contemporary Gabon, then, was an ominous premonition for Equatoguineans, a haunting glint that *this has all happened before*. Oil time is not only a transitory interlude but also a form of "serial and repetitive violent dispossession and appropriation, a sort of permanent frontier of exploration and abandonment" (Watts 2010: 10; see also Mason 2006).

On a road trip to Oyem with a group of Equatoguineans, several of whom had spent substantial parts of their childhood there but had not returned since, many shook their heads at the Soviet-style architecture, peeling paint, and visible structural damage that characterized so many of the buildings. They remembered it so differently—as new, modern, impressive. As younger people, this infrastructure had been imbued with the mystique and revulsion of nationalist exclusion that Moussa and other migrant labor now feel in Equatorial Guinea. Mateo, an Equatoguinean who had spent his childhood in Gabon and had returned to Equatorial Guinea to work for a U.S. oil company, lamented:

> I don't believe that the [Equatoguinean] state has learned. For example, the coastal promenade (paseo marítimo), which is very beautiful, is being extended. It has cost a fortune, but the value that it brings the population is infinitesimal compared to the value that could be produced by investment in the university to train technicians. This same concrete jungle went up in Gabon, in Nigeria, and it's the same thing we're doing here. There's concrete everywhere. The tendency is unstoppable: concrete, skyscraper, concrete, skyscraper. The only concrete that is good in my opinion is roads. All the rest—airports in the interior, promenades—it's all a waste. When you start investigating the wasteful use of resources, go to Gabon and look at the train.

Gazing at Oyem, the question of ruins begins to haunt the futurity of the present. Today Equatorial Guinea is dotted with the intricate woodwork and cathedral spires of Spanish colonial buildings in various states of restoration or disrepair. Small-scale cacao farmers wander the well-worn paths of once-large plantations, often living with their extended families in crumbling shells of former great houses, clotheslines stretching out into the sun. The infrastructures of oil time, similarly, present at least two related infrastructural landscapes that portend Equatorial Guinea's ruins of the future.

First, there are the infrastructures that have been at issue in this chapter—the asphalt roads, the paseo marítimo, the government palaces, even all of Oyala. Though they are designed to outlast the life cycle of hydrocarbon exploitation, just as their construction was funded by oil rents, so, too, are their maintenance and repair dependent on that income, unless the utopian dream of a diversified national economy—and a newly educated and financed local population—were to come to pass. Second, there is the corresponding landscape of potential ruins, more closely analogous to the Spanish colonial buildings, in the oil infrastructure itself—the refineries and storage plants, liquid natural-gas trains, walled residential and corporate enclaves, fixed offshore platforms. Oil ruins could be the new heritage industry, or the new sites that people might finally reclaim, incrementally, stringing laundry from pipe to pipe. As Fernando Coronil writes of Venezuela, "with the aging of the industry, abandoned oil camps and decaying oil towns scattered throughout the states of Zulia, Portuguesa, and Anzoátegui have added a ghostly aura to oil's former presence in Venezuela" (1997: 109; see also Burtynsky 2009; Effendi 2010).

For Phil, an American petroleum engineer living and working in Equatorial Guinea, the impending ruins there intertwined with his experiences of oil futures, now past, elsewhere:

> You always have abandonment and decommissioning issues, especially of oil- and gas-producing facilities. Eventually the resource runs out. The exact date isn't known.... [With Equatorial Guinea's gas-processing] facilities, the goal is to continue to use them long-term, but you have to find other sources.... So let's say in twenty or thirty years they will need to be abandoned and decommissioned.... In offshore you have to take infrastructure out: plug and seal wells, remove platforms. It's not an inexpensive endeavor. They make artificial reefs in the Gulf of Mexico, take out chemicals. Whether that's done here or not I don't know. Under most contracts, everything we buy at the end of the day belongs to the government. Until we finish, we can tear stuff down, sell it, et cetera. But once we're done we turn it over, it's theirs. The land is theirs. We turn it over to them.

The industry has "abandonment issues," which, by their nature, outlast the fact of the abandonment itself. Much of the offshore infrastructure—rigs, floating production, storage and offloading vessels—will move on. But other infrastructure that comes with oil and gas production will last long after the resource and its revenues have dried up, whether in "reefs" at the bottom of the ocean, abandoned plants with nothing left to process, or gated enclaves newly permeable to local use. Oil time—anomalous, cyclical, pregnant with abandonment—violently transects developmental time. From Oyem to sunken structures, oil's infrastructures are the ruins of the future.

Conclusion

When I lived in Equatorial Guinea I didn't complain about the running water. This is a shared reality that we all bear. But I had to complain that my country wasn't habitable by all. Electricity? We all share a difficult situation. But that not everyone is allowed to share even that situation? Unbearable.

———

The Equatoguinean writer-in-exile who spoke these words in 2009 reminds us of the dangers of a narrowly biopolitical, humanitarian concern that attention to infrastructure can summon: "By excluding the political, humanitarianism reproduces the isolation of bare life and hence the basis of sovereignty itself" (quoted in Robins 2009: 637). This is precisely the way that many Equatoguinean state actors used infrastructure to justify multiple deferrals—private infrastructures need to diversify the national economy before public infrastructures can provide goods and services; foreign contractors need to build this infrastructure before we can talk about access to credit for local businesses or contract law; public infrastructures need to be put in place before we can talk about political power, dictatorship, or repression. The writer's words are an admonition then about what is at stake in Equatorial Guinea's infrastructure development. Sporadic water and electricity are bearable, he reminds us, more so when they are a shared life condition. But that many openly dissident Equatoguineans (including the speaker) live in exile, unable even to share in the bearability of unreliable infrastructure, is to him unbearable.

And so I end with the question of infrastructure and political time, crosscutting, again, developmental time and oil time. Before oil, through the early 1990s, the formidable regime of Obiang (already in power for over a decade at that point) had been weakening. With external debts among the highest in Africa and conditionality-driven austerity wreaking havoc, there was a strong and growing opposition coalition that won an unprecedented number of seats in the 1992 parliamentary election. "And that's when petroleum started," explained an opposition politician. "Petroleum was like a life jacket for the regime, an oxygen balloon to help it float." With contracts signed and money newly pouring in, U.S. oil companies handed "the state" a strength and coherence it had all but lost. Today, Obiang is the longest-serving head of state in the world, having won reelection in April 2016 with 93 percent of the vote. In the image of petroleum as a life jacket for this regime, I would like to suggest that the opposition politician gives us another vision of infrastructure, a vision of unbearable lightness: infrastructure inflating seemingly overnight with the political oxygen of U.S. oil firms and the revenue and retrenchment of oppression they so reliably bring. Infrastructure's seemingly durable materialities—asphalt, concrete, steel, glass—are themselves a rigid materialization of Obiang's continued rule. Indeed, many if not most Equatoguineans would personally attribute (along a continuum of admiration to disgust) infrastructure projects to him—a ten-thousand-square-meter Italianate baroque basilica built at the cost of 9 billion CFA francs (~$17.5 million) in his natal town, and the new capital of Oyala where Obiang has demanded that entire new buildings be moved because he did not like the view, as extremes among innumerable examples. The construction of the memory of petroleum will also fundamentally be the memory of Obiang.

DEVELOPMENTAL TIME WAS ALWAYS a myth (Fanon 1963; Fabian 1983; Ferguson 1999; Chakrabarty 2007). Ethnographically, linear time fractures into constellations of futurity and deferral, teleology and stasis. But desire for the infrastructural world with which developmental time has come to be associated—paved roads and skyscrapers, running water and electricity—is a claim on material equality in a profoundly unequal world (Ferguson 2006). In Equatorial Guinea and elsewhere, the chimera of developmental "progress," infrastructural and otherwise, too often comes not through the reliable teleology of developmental time, but through the fitful temporalities of imperial formations, oil and development itself among them. The development industry is notoriously fickle, swayed this way and that by development trends. ("Did we say *Back to the Land*? Sorry, that was supposed to be *Back to the Factory!* All a big mistake" [Ferguson 1999: 241].) As Phil, the American petroleum engineer quoted earlier, demonstrates, the U.S. oil industry's presence in Equatorial Guinea is unambiguously temporary. Most simply, they will leave when the resource runs out. In other ways, however, the technological zone (Barry 2006) cum empire that is the transnational U.S. oil industry will linger long after the hydrocarbons are depleted. Like the cathedral spires of Spanish colonialism, the extensive infrastructures that oil and gas raise are the ruins and the rubble (Gordillo 2014) and the reefs of the future. While colonial administrators, contemporary Equatoguinean state officials, and many expatriate oil workers narrate these infrastructures in progressive terms, their constellation of deferral, promissory notes, abjection, and abandonment easily belie those narrations. "We are not independent yet," says my host family; "go to Gabon and look at the train" says Alberto. That the train built with petrodollars has not run in decades remains unsaid. Developmental time as *lived* has always been more like oil time—fitful and temporary, serial and cyclical. But the durability of Obiang's rule, indeed its seeming intractability testifies to the unbearable weight of oil extraction on political time in Equatorial Guinea and around the world. Under Anthropocene skies, petro-fueled infrastructures inflate like life jackets for a regime that otherwise should have drowned in its own violence and excess. As the promises and betrayals of developmental time, oil time, political time, and imperial time transect one another, the promise of infrastructure as an ethnographic category emerges. Attention to infrastructure enables new archaeologies of the present, multiscalar insights into the oscillating temporalities of today's imperial formations.

NOTES

1 Equatorial Guinea is widely ranked among the most corrupt and kleptocratic dictatorships in the world by Transparency International, the World Bank, Freedom House, and the Economist Intelligence Unit and other outfits who do this sort of thing. In 2016, Freedom House put the country in its "worst of the worst" category, along with North Korea, Sudan, and Turkmenistan. Obiang himself consistently places among the richest leaders in the world. Scholarly accounts too often frame Equatorial Guinea as exceptional. Even Bayart (1993)—who argues that "we should not draw too hasty conclusions about the privileged relationship between power and wealth [given that] the positions of power never absorb all the channels of wealth"—goes on to say that "only the political gangsterism of a Touré family in Guinea, or an Nguema family in Equatorial Guinea approaches a de facto confiscation of the means of wealth" (91). And yet the now approaching four-decade rule of Obiang Nguema Mbasogo is, like all state projects, a "multilayered, contradictory, translocal ensemble of institutions, practices, and people" (Sharma and Gupta 2006: 6). While Obiang, his family members, and their close associates have long benefited from absolute rule and access to positions of power, it is also important to stress the fractures and tensions at the very heart of the regime.

2 National Economic Conference II, National Plan for Economic Development, Agenda for the Diversification of Sources of Growth, Equatorial Guinea toward Horizon 2020. In 1997, Equatorial Guinea held its first national economic conference. There had been national-level conferences before, but the 1997 conference was distinguished by the fact that it was conceived, documented, and publicized by its state organizers as Equatorial Guinea's "first" national economic conference (Appel 2017). Oil had been discovered three years earlier by an independent American oil company, and small amounts of money from exploratory contracts and field leases had

just begun to circulate back into Equatorial Guinea. This was a dramatic turn of events for a microstate characterized by unprecedented economic collapse in the late 1970s, creeping into a crippling debt burden by the early 1990s.

REFERENCES

Appel, Hannah. 2012a. Offshore work: Oil, modularity, and the how of capitalism in Equatorial Guinea. *American Ethnologist* 39(4): 692–709.

———. 2012b. Walls and white elephants: Oil extraction, responsibility, and infrastructural violence in Equatorial Guinea. *Ethnography* 13: 439–465.

———. 2017. Toward an ethnography of the national economy. *Cultural Anthropology* 32(2): 294–322.

Appel, Hannah and Mukul Kumar. 2015. Finance. Theorizing the contemporary, *Cultural Anthropology* website, September 24, 2015. https://culanth.org/fieldsights/717-finance.

Barry, Andrew. 2006 Technological zones. *European Journal of Social Theory* 9(2): 239–253.

Bayart, Jean F. 1993. *The state in Africa: The politics of the belly*. Hoboken, NJ: Wiley Press.

Benjamin, Walter. 2002. *The arcades project*, ed. Rolf Tiedemann. Cambridge, MA: Harvard University Press.

Burtynsky, Edward. 2009. *Oil*. Gottingen: Steidl.

Carse, Ashley. 2014. *Beyond the big ditch: Politics, ecology, and infrastructure at the Panama Canal*. Cambridge, MA: MIT Press.

Chakrabarty, Dipesh. 2007. *Provincializing Europe: Postcolonial thought and historical difference*. Princeton, NJ: Princeton University Press.

Coronil, Fernando. 1997. *The magical state: Nature, money, and modernity in Venezuela*. Chicago: University of Chicago Press.

Effendi, Rena. 2010. *Pipe dreams: A chronicle of lives along the pipeline*. Amsterdam: Schilt.

Engels, Frederick. (1884) 2010. The origin of the family, private property, and the state. Marxists Internet Archive. https://www.marxists.org/archive/marx/works/download/pdf/origin_family.pdf.

Fabian, Johannes. 1983. *Time and the other: How anthropology makes its object*. New York: Columbia University Press.

Fanon, Franz. 1963. *The wretched of the earth*. New York: Grove Press.

Fegley, Randall. 1991. *Equatorial Guinea*. Oxford: Clio.

Ferguson, James. 1999. *Expectations of modernity: Myths and meanings of urban life on the Zambian copperbelt*. Berkeley: University of California Press.

———. 2006. *Global shadows: Africa and the neoliberal world order*. Durham: Duke University Press.

Gordillo, Gaston. 2014. *Rubble: The afterlife of destruction*. Durham: Duke University Press.

Guyer, Jane. 2007. Prophecy and the near future: Thoughts on macroeconomic, evangelical, and punctuated time. *American Ethnologist* 34(3): 409–421.

Harrison, Virginia. 2013. World's best economies in 2013. CNN. http://money.cnn.com/gallery/news/economy/2013/12/27/best-economies/2.html.

International Monetary Fund. 2009. Republic of Equatorial Guinea: Selected issues. IMF Country Report No. 09/99. Washington, DC.

Kennel, James et al. 2009. The emancipatory politics of the dialectical image. *Reading the arcades / reading the promenades*. https://arcadespromenades.wordpress.com/tag/convolute-n.

Larkin, Brian. 2013. The politics and poetics of infrastructure. *Annual Review of Anthropology* 42: 327–343.

Limbert, Mandana. 2010. *In the time of oil: Piety, memory, and social life in an Omani town.* Stanford, CA: Stanford University Press.

Mason, Arthur. 2006. Images of the energy future. *Environmental Research Letters* 1(1): 1–4. https://doi.org/10.1088/1748-9326/1/1/014002.

Mba Abogo, César A. 2011. *(La construcción de) la memoria del petróleo*, ed. Dulcinea Tomás Cámara. Alicante: Biblioteca Virtual Miguel de Cervantes. http://www.cervantesvirtual.com/obra/la-construccion-de-la-memoria-del-petroleo.

Ministry of Mines, Industry, and Energy. 2008. Production Sharing Contract (sample). Republic of Equatorial Guinea MMIE website. http://www.equatorialoil.com/PDFs%20for%20download/Model%20PSC_2006_English.pdf.

Morgan, Lewis Henry. 1877. *Ancient society.* New York: Henry Holt.

Mrázek, Rudolf. 2002. *Engineers of happy land: Technology and nationalism in a colony.* Princeton, NJ: Princeton University Press.

Robins, Steven. 2009. Humanitarian aid beyond "bare survival": Social movement responses to xenophobic violence in South Africa. *American Ethnologist* 36(4): 637–650.

Roitman, Janet, and Achille Mbembe. 1995. Figures of the subject in times of crisis. *Public Culture* 7: 323–352.

Rostow, Walter W. 1960. *The stages of economic growth.* Cambridge: Cambridge University Press.

Sharma, Aradhana and Akhil Gupta, eds. 2006. *The anthropology of the state: A reader.* Hoboken, NJ: Wiley Blackwell.

Stoler, Ann Laura. 2008. Imperial debris. *Cultural Anthropology* 23(2): 191–219.

Watts, Michael. 2004. Violent environments: Petroleum conflict and the political ecology of rule in the Niger Delta, Nigeria. In *Liberation ecologies*, 2nd ed., ed. Richard Peet and Michael Watts, 273–298. New York: Routledge.

———. 2010. Regimes of living: Life and death on the Nigerian oil fields. Paper delivered to Capitalism's Crises workshop at Stanford University, April 13.

The Future in Ruins: Thoughts on
the Temporality of Infrastructure

AKHIL GUPTA

This chapter explores the temporality of infrastructure by focusing on the relationship between futurity and ruination. While the relationship between infrastructures and the future has long been appreciated, this gesturing to the future seldom attends to the fact that infrastructures are always already on the way to becoming ruins. Ruination draws our attention to the ongoing work of maintenance that is necessary in order to give infrastructure projects their appearance of solidity and immovability (Carse 2014; Barnes 2016). I will argue for a view that looks at infrastructure as an open-ended process rather than through the teleologies of "completion" and "progress" (Graham and McFarlane 2014). Briefly, my argument is that the conventional view of infrastructural projects as beginning with planning and ending with inauguration misses the dynamic nature of infrastructural time in favor of a well-worn script of modernity. I will advance a contrarian view that sees infrastructures as a process that is characterized by multiple temporalities, open futures, and the constant presence of decay and ruination. I conceptualize infrastructure as a process, not a thing: a thing-in-motion, ephemeral, shifting, elusive, decaying, degrading, becoming a ruin but for the routines of repair, replacement, and restoration (or in spite of them). If one adds to such a complex temporality the peculiar features of infrastructure in the global South, we have an even more striking calling into question of the implicit teleologies that mark infrastructural time (Barak 2013). I explore here this rich and contradictory set of relations between infrastructures and their futures.

Infrastructures have always been considered forward-looking for a number of reasons. Most infrastructural investment tends to be "lumpy": it involves large capital outlays as well as significant expenditures of other resources, such as land, energy, or human resources. Such investments typically pay off over a long period of time, ranging from a few years to fifty years or more. Thus, investments in infrastructure always involve calculations about the future, and because infrastructures are usually public goods, the calculation of return is often uncertain. The hope is that building infrastructure will result in higher levels of economic growth and improvements in the well-being of the population. In this sense, investments in infrastructure are intended to bring about a desirable future but whether that future will actually come to pass is always unpredictable (Koselleck 2004).

Borrowing from Arjun Appadurai's (2013) work on the future, we can say that infrastructures tell us about aspirations, anticipations, and imaginations of the future: what people think their society should be like, what they want it to be like, and what kind of statement they wish to make about that vision of the future. Infrastructures are concrete instantiations of visions of the future.

Another important role played by infrastructure is in shaping the present through a politics of anticipation. Perhaps because of the investment involved, infrastructure is almost always built to exceed present needs: it is built in anticipation of a not-yet-achieved future. For example, a new airport terminal may be built not for present levels of traffic, but to handle the traffic anticipated in ten years. This anticipatory side to infrastructure means that it is always future-oriented, and the horizon of anticipation might be anywhere from a few years to long into the future. For this reason, perhaps more than most social phenomena, infrastructure makes clear how the future configures the present.

It turns out that we can understand a great deal about social futures by looking at infrastructure. We know that infrastructures fix space and time because, once finished, they are hard to reverse. For example, a highway cleaving a community might end up destroying it, but even if most people realize that the highway is a mistake, it is unlikely to be dug up or moved somewhere else. The reason lies not only in the capital investment required to reconstruct the highway, but also because the location of the highway generates its own infrastructural effects in the location of new houses, schools, or businesses.

There are other reasons for thinking about infrastructure as a technology of and for the future that go beyond its role in fixing time and space. Infrastructure should be seen as a concrete metaphor: as a physical presence in the landscape that channels communication, travel, and the transportation of goods; a biopolitical project that aims to address the health and welfare of the population while also facilitating discipline and control; and an aspirational project that functions as the symbol and index of a future becoming (Harvey and Knox 2015). The latter two aspects of infrastructure are as important for its role in shaping futures as the first. Focusing on the relationship between the biopolitical and the infrastructural, the government of people and of things, may help us better formulate an anthropology of the future and of the present.

Another way to link these three aspects of infrastructure is to emphasize how infrastructure shapes the biopolitical through its role in imaginations of the future. If one looks at the imagination of national futures, infrastructure has been a key mechanism by which biopolitical and political imaginaries have been constructed. Infrastructure represents that which makes a nation-state modern and developed. State-of-the-art infrastructure represents a modern nation-state. One of the most famous statements in the history of postcolonial India, one that every schoolchild in India can narrate, is a quotation from Jawaharlal Nehru, India's first prime minister, when inaugurating the Bhakra Nangal dam in Punjab: "These are the temples of modern India." With his keen sense of history, and his unique ability to articulate the promise of modernity, Nehru fused the past with the future and infused the infrastructural with spiritual affect.

One can see the rush to build highly visible infrastructure in Third World countries that are becoming wealthier as explainable not simply by their growing needs for such infrastructure, which are no doubt real, but also by what such infrastructure represents: that the nation-state is now modern, industrial, and developed. When China held the Olympics, people from the rest of the world marveled as much at the amazing infrastructure that had been put into place as they did the Games themselves. Indeed, in the race to be counted as a developed nation-state, infrastructure is critical. Nothing announces that a place is a world-class city or a developed nation-state as much as its infrastructure (Anand 2006). If the infrastructure in Shanghai makes the West look "backward" by comparison, it is only because we have completely internalized an "infrastructure clock" in which cities, regions, nations, and even continents can be organized into backward and forward, ahead and behind, and developed and undeveloped. Allochronism is instantiated through the state of infrastructure.

Another way in which the biopolitical and the infrastructural are linked is through bureaucratic mechanisms. The argument about "new public goods," such as transparency, accountability, power sharing, and offering choice to consumers, is that by operating on the institutional and bureaucratic mechanisms that regulate infrastructure, they enable more efficient delivery of those products (Bear and Mathur 2015). The argument for accountability is never going to be that it builds roads, but, for the roads that are built, it hopes to make the project cheaper, more efficient, and more equitable. If the debate on infrastructure so far focused on the political and social consequences of things like the displacement of people by big dams, the purchase of land from farmers under eminent domain, and so on, then the new public goods draw our attention much more to institutional reform, to the bureaucracies that are a necessary part of infrastructural development. They derive from the recognition that the delivery of the old public goods needed bureaucracies, and that these bureaucratic structures were as important to the provision of infrastructure as the cement and other materials needed for construction. Tempting as it is for us to go that route, we should not treat bureaucratic structures as a "secondary" feature of public works. As Marx reminds us constantly throughout *Capital*, the organizational forms in which production takes place are as much a part of the production process as the tools and machinery. In fact, when institutional and organizational forms are referred to as the machinery of production, the usage is not analogical, but literal. A production system is not only constituted by a number of lathes, but by their ordering in rows, with division of tasks for each operator. The analogy I wish to draw is that a similar logic operates in the case of bureaucracies and public goods.

The "new" public goods, therefore, enable us to make explicit the link between the biopolitical and the infrastructural, and this may constitute a good place to think about the role of infrastructure in modernity. Focusing on the bureaucracies that enable infrastructures to be built and operated also helps see infrastructures as a process, an operating assemblage that requires constant work, rather than as a static object.

The Future as Ruins

I now consider the future as rubble and the ruins of the future. In formulating it thus, I am not engaging merely in metaphorical play; I intend to tie it concretely to the infrastructures of modernity (see also Schwenkel 2013). That "concrete" is so often used as a synonym for "material" is no doubt a reflection of nineteenth-century infrastructure; the materiality of radio waves and bandwidth may inspire other metaphors and other substances, or perhaps their materiality will disappear in the metaphors in which we will talk about the infrastructure of the future.

The three dimensions of infrastructure highlighted above—a channel that enables communication, travel, and the transportation of goods; a biopolitical project to maximize the health and welfare of the population at the same time as subjecting it to control and discipline; and its role as the symbol of a future being brought into fruition—are especially visible in the energy sector. This is not surprising, because the political and social significance of energy is rising as the problem of climate change becomes clearer. But there are other reasons as well. As large and formerly poor countries in the world like China and India industrialize, their demand for energy resources and infrastructure is shooting up, from coal and oil to solar and wind energy. The social implications of this demand are enormous, from mass displacement of populations to environmental consequences.

That energy politics is a critical axis of biopolitics becomes evident when we understand that the management of the population requires the expenditure and regulation of sources of energy. Whether it is food supplies or oil resources, controlling and accessing these sources of energy have been critical to states' management of their populations (Collier 2011). Yet very little of the literature on biopolitics has engaged with sources of energy. The idea of public goods brings together infrastructural concerns about food, water, electricity, and energy with the biopolitical objectives of increasing the well-being of the population. Governmental provisioning of infrastructure (public goods seen as "natural" monopolies) is an essential component to maximizing the welfare of the population. Infrastructures for provisioning the population with food, water, energy, healthcare, and schooling, and of providing transportation and communication, are often considered *necessary* for bringing about an acceptable level of development. Increasingly, these infrastructures are being privatized and rebranded as PPPs (public-private partnerships).

One of the long-standing nationalist critiques of colonial infrastructure was that it was specifically intended to bring about a particular kind of future that was ruinous for the colonized nation-state. Nationalists and other anticolonialists charged that colonial infrastructure such as railway lines and ports were often intended to better extract raw materials and to enslave and indenture laborers for the benefit of colonizers, rather than to improve the lives of people in the colonies (Rodney 1981). Here we see the clearest statement about how the economics of infrastructure is always tied to its politics. Infrastructures are important because the future they bring about always favors one set of political actors over others. There is no such thing as politically neutral infrastructure.

In a rapidly urbanizing world, we should remember that urban dwellers are more dependent than their rural counterparts on infrastructures that deliver to them food, water, electricity, natural gas, gasoline, and sanitation. In most settings in the global North, infrastructures disappear from view: pipes delivering water and providing sanitation are hidden in walls, electricity is delivered by an outlet, and natural gas heats up stoves and ovens by turning a valve.

However, the biopolitical project embodied in infrastructure fails when infrastructure breaks down: the "normal" condition of infrastructure in postcolonial settings is one where infrastructures function intermittently, if at all. Against the backdrop of 24/7 supply, electricity and water become "problems" that are noticed only when their supply is restricted to a few hours in the day or not at all. But how does one think about the "normal" when people usually get electricity or water for a few hours in a day and for a few days in a week? Such situations are not anomalous and make infrastructure a good means to think of temporality and futurity. I will take up one example of such a failure of infrastructure to argue that the actual practice of building infrastructure may end up subverting the imagined future that is embedded in an infrastructural project.

The pedagogical and performative role played by infrastructure in the biopolitical project of managing and controlling the population of a nation-state and in the political project of interstate competition has not been fully appreciated (Bhabha 2004).[1] Some examples will help make this point. One of the most interesting, and least observed, facets of postcolonial Indian cinema has been the representation of big infrastructure and its explicit yoking to the future of the nation-state. Various state governments measure their success in providing "development" with the provision of electricity to villages. As the time approaches for an election, the ruling regime will take out full-page advertisements in all newspapers detailing the number of new villages it has "electrified" (the official term, "electrified," itself conveys the jolt of modernity). Iconic films like *Mother India* cemented the relationship between large dams and the provision of irrigation water and electricity for poor farmers. Only the large infrastructural projects being initiated by the postcolonial state, it was suggested, held the promise of freeing poor farmers from poverty.

Perhaps nothing better demonstrates what infrastructure signifies than the fact that these projects themselves are constructed as tourist destinations. The pedagogical and performative roles played by big dams are seen most clearly by the planning that goes into deciding how tourists will encounter the dam, which parts they will see, how many people will be allowed to enter, how large the parking lot should be, and so on. In India, I have often gone to family picnics at these large dam sites: they are built with beautiful landscaping, lawns, flowering gardens, and high-tech fountains so that people can come and participate in the making of the modern nation-state. If one looks at any "modern" infrastructural project in India, such as nuclear power plants, high-tech research institutes, dams, and airports, they are always surrounded by beautiful green lawns and flowerbeds, even when the facilities are off-limits to visitors. What these sites represent is a certain kind of modernist aesthetic, one in which function is allied to a particular idea of order and beauty. They represent a vision of the future, and the grounds themselves instruct the people admitted into them about how to inhabit such a future.

I returned to the village in north India where I did my fieldwork and found many people abuzz with this encounter with the modern. There was a new government dairy that had opened in the area, which was now buying most of the milk produced in the village. Everybody who sold milk to the dairy was bussed free of charge to the plant, where they were given a tour of what happened to the milk once it left their hands. Villagers told me with great astonishment about how the milk was never touched by hand again, but instead went through a whole array of machines to come out in cartons at the other end. The pedagogical goal served by the "visibility" of infrastructure here is not so much about teaching hygiene to villagers but about the biopolitical project of creating citizens who share the goal of inhabiting a modern future.

Suspension

To anticipate the argument about the nature of infrastructural time, I will look at examples of infrastructure in the global South and use them to think about one particular type of temporality that I call "suspension." I will ask: What kinds of futures are implicit in the idea of suspension? How does one theorize the future from suspension? What is the relation between suspension and ruination? When one recasts infrastructures as always being in motion, then suspension emerges as one particularly interesting modality of infrastructural temporality.

In his study of ruins, Gaston Gordillo (2014) deals with what he terms "the afterlife of destruction"; my concern is more with the afterlife of construction. Specifically, I am interested in the role that infrastructure plays as an index of modernity and symbol of development. Like many other cities in India, Bangalore aspires to be a "global city" that is part of a rising nation-state. Bangalore's aspiration to be a global city, however, is even more keenly felt, and strongly held, by its citizens because it has become the technology hub of India, at least in its leading sector, Information Technology (IT) and IT-Enabled Services (ITES), which includes the large Business-Process Management (BPM) sector. Traversing the city's streets is a nightmare, not only because there are too many cars for the narrow roads, but because the roads are always dug up for one infrastructure project or another.

There are new pipes being laid everywhere, pipes carrying water, sewage, and natural gas, of course, but also pipes enclosing electric wires, high-speed Internet cables, and telephone cables. Roads are also being dug up because flyovers are being constructed above them, or underpasses below, and because the roads themselves are being widened. Roads are also dug up because a raised platform for the metro is being built in the middle of some of the busiest roads. To the city's north, a raised expressway that provides a "fast link" between the new airport and the city was under construction for many years, and another raised expressway has been constructed between the city and Electronic City in the south, where many IT companies are located.

Many of these sites are left unfinished for months and years on end, as if they had been absentmindedly started and then abandoned when the construction company got a new contract somewhere else. A visitor from another planet who did not know about these projects' embeddedness in a particular teleology—that they were part of a project to make the IT capital of India into a world-class city—would find them to look exactly like ruins. They could very well be part of a narrative that spoke of a place that once had elevated highways, and elevated metro lines, and flyovers, but whose infrastructure had turned into ruins.

Surrounding these ruins is rubble, alongside and on the roads: broken pieces of brick, tangled rebar, and broken concrete panels; kachā roads made of mud and stone; huge patches of tar and stone that have been used to hastily refill holes; pipes of all sizes and shapes, some of them broken and cracked; piles of dirt and rock; and dust, the particular dust of destruction, the rubble that emerges when buildings are demolished or foundations excavated.[2] Driving in Bangalore is to maneuver around this rubble, to negotiate the bumps and slowdowns it creates, and to aspire to a better future, to a future with "world-class" infrastructure.

This is a form of ruination with a very particular temporal structure: they are the ruins not of the past, but of the future. One contemplates ruins with nostalgia and associates them with degeneration and decline, with the elegiac decay of once-flourishing civilizations. Ruins are normally seen to be at the end of a historical trajectory that may be periodized as ascent, flourishing, and decline, what may be called a civilization's postsenescence. It is not surprising that often ruins are precisely those large infrastructure projects that in their heyday heralded the rise of a civilization. Thus, we have ruins of canals and aqueducts, and the broken remains of bridges, public buildings, highway rest stops, or the stone edicts of Ashoka. There are also, of course, the ruins of monuments that were built to represent the grandeur of their civilizations, to make their spectators awestruck. Very few infrastructural projects were built simply for their function. One can think of ruins, thus, as representing the afterlife of infrastructure.

The ruins I speak about—the ruins of the future—are inserted into a very different temporal structure. It makes sense not to think of them as occupying a temporary zone between the start of projects and their completion. Every pillar sticking out of the middle of a road marks the temporality of the now, between past and future, between potential and actualization. Ruination is not about the fall from past glory but this property of in-between-ness, between the hopes of modernity and progress embodied in the start of construction, and the suspension of those hopes in the half-built structure. Rubble here stands neither for senescence nor for anticipation, but for the suspension between what was promised and what will actually be delivered. To label them "incomplete" would be to succumb to the narrative of completion as telos; we should think of this suspension, rather, as a condition in its own right, not a transient property on the way to becoming something else. Rather than theorize it as immanence, the emergent suspended highway already visible in its pillars, I see such ruination as its own condition, its own end.

Theorizing the Future from Suspension

The half-complete infrastructural projects that seem to litter the landscape in Bangalore cannot be theorized as a temporary condition, a short period in which the rubble of construction—materials not yet used, and materials that have been destroyed—occupies the roads, to be swept away when construction is complete. In-between-ness, this space between past and future, is produced by the structural conditions of building infrastructure in India. However, my point is that such a condition is not at all unusual and may be characteristic of large infrastructure projects all around the world. A view of infrastructure that emphasizes its ongoing and processual nature makes suspension one of many possible trajectories, rather than as a unique break or interruption of a teleological timeline resulting in a finished object.

In India, as in many parts of the world, large projects are often stuck in their implementation. Even if they are eventually "completed," big infrastructure projects are often suspended due to legal challenges. The important point about these delays is that they are unpredictable. Once started, a project may eventually be completed, abandoned, or destroyed. That is why one must avoid the teleology implicit in terms such as "delay" and "completion": that is just one possible trajectory of infrastructure. Most often, legal challenges are due to the confiscation of private land under eminent domain for the purpose of building infrastructure. In such cases, the courts issue a "stay order," prohibiting construction until the legal issue is resolved. When there are many parties offended by land confiscation, the process can take an unpredictably long time to clear the courts.

But legal challenges are not the only reason so many projects are either abandoned or eventually destroyed. Capital shortages plague infrastructure projects in India. The regional state or municipality starts construction with great enthusiasm, but eventually the money runs out, and the project lies suspended until the next tranche of money is released. Waiting for money is fraught with its own uncertainties. A project may lose momentum because a politician who supported that particular infrastructure is voted out of office in the next election.

Then there are other kinds of money pressures. The politicians and bureaucrats who authorize the approval of government funds for a project typically keep a large chunk for themselves.[3] If a company pays off the amount required, and the government, ministers, and functionaries change because of an election, the company may balk at paying a new set of administrators and politicians. This sets up conflict between the authorizing agency and the construction company. The duration and intensity of this conflict are unknowable, although both parties have an interest in keeping the resources flowing for construction.

Rather than imagining incomplete infrastructure, or infrastructure in the process of construction, as always already on the path to completion, we should think of the end as potentially open. This is as much an empirical point as it is a philosophical one, for we often hear about projects that are begun but never finished. Infrastructure projects can be delayed for small or long periods of time before they are completed, or they can be suspended by being abandoned, or they can be reversed by active destruction.

Take the example of Narita Airport, the largest international gateway into and out of Japan. It was mired in controversy from the very beginning and has still not achieved the capacity for which it was planned in the 1960s. Its location was decided without consulting the communities to be affected because imperial land was available. But additional land had to be acquired, and a vigorous resistance movement soon developed, escalating to armed conflict, land occupation, and the building of structures to impede the construction of the runways. Eventually militants took over the control tower and smashed most of the equipment inside. The airport, initially scheduled to open in 1972 with five runways, was finally inaugurated in 1978 with a single runway and high security throughout. We can say, thus, that Narita Airport between the mid-1960s and 1978 was in a state of suspension. Although we can narrate its history in a manner that saw its completion as inevitable only retrospectively, such Whiggish narratives dominate the understanding of infrastructure. In fact, they are so deeply entrenched that when infrastructures are abandoned or suspended, we implicitly define that as an anomalous situation, rather than empirically inquiring whether such a state of affairs is unusual.

If we look at infrastructure projects neither from the perspective of the neat charts and timelines of planning documents nor from the retrospective view afforded by the cutting of the tape at their inauguration, but from the time when construction is under way and perhaps making uncertain "progress," we would perhaps see something that eludes us otherwise. As indicated above for the Indian case, construction may be halted for many reasons: there may be engineering problems; social pressures due to political opposition may reach a head; unforeseen environmental issues may crop up; money may run out; courts may halt proceedings due to lawsuits; governments might change; and the construction company may go out of business or run into financial difficulties. The path of any large infrastructural project seldom runs smoothly, and it seldom follows the timelines laid out in planning documents.

The time of suspension, of the hiatus, of the pause, is also a time of relative temporal openness. The future is unknown and unknowable: the project may go ahead, or be scrapped, abandoned, or modified. I argue that this openness to different outcomes in the temporality of infrastructure is important to preserve because it allows us to tell richer narratives about the life of infrastructure.

In India, ruination prefigures even the completion of projects. Inferior construction, partly caused by the use of substandard materials that companies are forced to use after paying such a large share of the budget as bribes, prefigures infrastructural projects as ruins even before they are complete. Ruins can thus be said to be immanent in the completion of projects. As soon as the project is complete, and officially declared to be open, it starts being repaired. This process of "repair" is fueled by the allocation of annual expenditures on leaky infrastructure, of which another good share is siphoned off by politicians and bureaucrats. Ruination prefigures and configures infrastructure that is "completed": the ruins of the future portend the future as ruins.

Shifting Infrastructure: The Process of Ruination

Massive infrastructural projects like roads, dams, and airports invoke images and ideas that are diametrically opposed to notions of the ephemeral, the shifting, and the elusive. The typical life stages of these massive infrastructure projects are differentiated very clearly at the front end. Controversies often surround the location of airports, focusing particularly on land acquisition, transportation links, and environmental costs. Similar controversies have surrounded roads and dams, as evidenced from the controversy surrounding the Sardar Sarovar dam on the Narmada River in India, where the chief issues were the displacement of indigenous peoples and the ecological damage caused by the dam. Large infrastructure projects that require massive amounts of investment and resources (concrete) thus seem to follow a peculiar trajectory: they are contested from the time that they are conceived to the time that they are "completed." After that, controversy usually deserts them: they are forgotten precisely because they are functioning normally. They appear to disappear when they do what they were supposed to be doing in moving people, commodities, water, electricity, gas, or oil. In other words, completion is a form of death for these projects not only because they cease to be objects of controversy but also because they seem to disappear at the very moment that they become socially useful and actually begin to fulfill their social role.

These projects are in the limelight in the phase of "becoming" but disappear from social scrutiny, discussion, and debate in the phase of "being." Why is that the case? What would happen if we adopted a very different way of understanding infrastructure in which the start of a project did not coincide with its planning and the end with its completion? Is there a different way to imagine the structure of the temporality of infrastructure in which the telos was not completion? In other words, would other ways to think the temporality of infrastructure help us move beyond an endpoint that is really the moment when the infrastructure begins operating?

One can take a very different look at such projects, invoking ideas of decay, ruination, repair, and the constant work required to make these projects icons of stability and progress. Infrastructures are often imagined by the public in static terms, corresponding to the idea that infrastructures work in the background and do not come to our attention unless they cease to function. Once finished, infrastructures occupy a dead time, an inertial existence, until they break down and are suddenly thrust into the temporality of birth, life, and decay. Such a static view of infrastructure also renders the often long periods of time when they are being "constructed" problematic, because they have been started but are not yet complete.

By contrast, a dynamic view of infrastructure enables us to replace the social death of a project marked by its completion with a focus on movement and process, on the constant struggle between renewal and ruination. Such an emphasis allows us to theorize infrastructure as perpetually in motion, always shifting and ephemeral, and often elusive. Beginning from movement requires a completely different optic, making us think of periods when infrastructures are "at rest" as not the normal condition, but as something to be explained. Has movement truly been arrested? Or does it merely *appear* as if movement has been brought to a halt, and, if so, why?

It is precisely the appearance of solidity that makes it hard for us to think of infrastructures as always in motion. We think of highways, metro stations, airports, and power plants as solid, immovable objects, and this is exactly what enables infrastructures to disappear from consciousness. The highway, the metro station, or the airport may enable circulation, it may become a critical node of circulation, but in doing so, it disappears from circulation itself. It enables people and commodities to move, but it does not itself appear to move.

A focus on infrastructures as emergent, always in process, always shifting, changing, decaying, being rebuilt, and being maintained, not only excavates the labor involved in building and maintaining infrastructures, but also draws attention to the properties of the materials themselves and to the representational work done to and by them. Coming back to the case of Narita Airport, it was inaugurated almost a decade later than planned under heavy security as fourteen thousand riot police kept out several thousand protestors. This security remains up to the present as a constant reminder that the work of securing the airport is continuous and ongoing and did not end once the airport began operations and the protests subsided. Also, Narita was only built with one runway instead of the five originally planned, which had already been scaled down to three after initial protests in 1965. A second runway was not built until the World Cup in 2002, and it did not allow for normal operations until 2011 because opponents had built a three-story building in its path. Farmers and homeowners refused to sell land to the government to expand the second runway. A sixty-five-year-old farmer explained why he refused to sell his land despite not being able to hear anything from the noise of overhead planes. "If they [the airport's operator] had cared about the people who are living here, the situation would have turned out differently," the farmer said. Local officials were sympathetic to the people whose land the airport wished to purchase (and eventually usurped by decree). One official complained that the airport "may be moving toward completion in terms of its utility, but it lacks soul." In 2012, an official at the airport said, "As some land has yet to be acquired, we cannot say the airport is complete. We would like to continue dialogue" with stakeholders (*Japan Times,* April 24, 2012).

The example of Narita points to something important about large infrastructural projects: they are always in process. Almost forty years after it was completed and officially opened, Narita continues to strive toward completion. The second of three runways has not yet been completed and the planned third runway is nowhere in sight. A train line linking the airport to Tokyo was not built until 1991 and a high-speed rail link was not available until 2010.

Nor is such a situation peculiar to Narita: Heathrow Airport in London has been the subject of a long-running controversy about the addition of a third runway and a new terminal building that has pitted local residents against the British government and the airport authority. In some airports, slots for expansion are built into the initial plans but in a case like Heathrow, the existing infrastructure has been rendered insufficient because demand has risen beyond anything projected at the time when the airport was planned.

If infrastructures are always incomplete, always in process, the inauguration of an infrastructural project that announces its completion is therefore always an ideological act (see also Harvey and Knox 2015). What does it mean to say that a project is "complete" and ready for use when only a third of the plans have been executed? Who decides that the project is complete? What does it mean to represent a project as being complete? Who is the "public" conjured by these representations?

The raising of funds for an infrastructural project involves a different temporality from construction and use. Large projects require huge sums of money and may need to be sold to a skeptical public. "Selling" a project may be literal, in that government bonds may be required, or it could be metaphorical, in the sense of obtaining legitimacy for the use of public funds. A larger share of the budget of big projects such as highways and airports is now routinely set aside for public relations and for mitigating the impacts on those directly affected. For example, airports now offer to replace the windows of homeowners in the flight path with new sound-proof, double-insulated windows. The Big Dig in Boston had a large budget set aside for "mitigation," essentially paying off anyone who was inconvenienced by the construction. Public relations campaigns are launched to make sure there is political backing from the public; legislators are lobbied to make sure there is support in the legislature.

The "selling" of a project may be solely for the purpose of garnering legitimacy. Very often, in the global South, projects are sold by the promise of modernity that they offer. A high-speed rail line, a new airport, a highway connecting major cities, and new museums and other civic buildings are all legitimized by appealing not to their utility, but to their (much more intangible) promise of bringing modernity. These projects then function by doing something that does not involve moving people and things, but by creating forms of national and regional identity. Those who live in a world-class city may do nothing to improve the life of slum dwellers, but they may well take pride in its shiny new buildings or the new subway line, even when it leads to their own displacement (Ghertner 2015).

Consciously seeking legitimacy for infrastructure projects may enable expansion and completion-in-stages. But these are only two modalities by which infrastructures change. By far the most important factor in why infrastructures change is the decay and degradation of the materials used in construction. Infrastructures require constant work to function. This labor, which involves the replacement and repair of broken or dysfunctional parts, is often glossed as "routine maintenance." And it is precisely the routines of repair, replacement, and restoration that give infrastructures their appearance of solidity, of sturdy immobility.

Another important reason why infrastructures change is because of technological obsolescence. This rarely happens due to the "disruptive technologies" so beloved by Silicon Valley, but by small, cumulative changes that require continuous "improvements" to be made in existing infrastructure. Anybody involved in overseeing the functioning of a large infrastructural project knows that the job never simply ends when construction is complete. Rather, what happens is that one set of tasks ends and a whole other set begins. The two tasks may not be identical in their challenges or in their public appeal, but they are equally important, for without the work of maintenance, the work of construction does not last.

By conceptualizing ruination not as that which attacks infrastructure *after* its lifetime, but as something that is endemic to infrastructure in the form of rusting and leaking pipes, degrading cables, equipment ruined by water leaks, bolts becoming loose, and roads developing potholes, I am drawing attention to the temporality of the materials used in making infrastructure. Because these materials have their own cycles of decay and degradation, from the moment that construction is "complete," ruination starts operating. What keeps infrastructure functioning, then, is the continuous work of maintenance. We can thus see how infrastructure lies at the intersection of different temporalities: human and nonhuman, social and technical, and material and ideological.

Because infrastructure has multiple temporalities, its future is open. The effects desired by the investment in infrastructure in the form of improved well-being of the population may come about or not. More likely, it may come about with unintended effects: the modern city with good roads may encourage the operation of so many cars that air pollution may rise to dangerous levels, as has been the case with Beijing and Delhi. The multiple futures of infrastructure arise not only because of the decay and degradation of materials, but also because of the complex interplay between its social, technical, material, ideo-

logical, performative, pedagogical, and aspirational temporalities. By conceptualizing infrastructures as dynamic things-in-motion, as mobile assemblages,
we can better grasp the relationship between futurity and ruination that lies at
the heart of infrastructural time.

NOTES

I would like to thank Sumita Mitra for research assistance. I am very grateful to audiences
in the following places who gave me useful comments on different parts of this essay:
"The Anthropology of Infrastructure" panel, 112th Annual Meeting of the American
Anthropological Association, Chicago, November 2013; International Institute of Asian
Studies' Macau Winter School, December 2013; Center for Energy and Environmental
Research in the Human Sciences (CENHS), Rice University, March 2014; Heidelberg
Summer School 2014; School for Advanced Research, Santa Fe, November 2014; The
Tod Spieker Colloquium, UCLA Department of Geography, April 2015; University of
Melbourne, August 2016; and, "Cultures Of/As the Ephemeral, the Shifting, and the
Elusive," 115th Annual Meeting of the American Anthropological Association, Min-
neapolis, November 2016.

1 Bhabha (2004) is dealing with nationalist affect, not with infrastructure in particular.
2 *Kachā* literally means "unripe." In the context of roads, it is a reference to gravel roads,
  that is, roads that have not been macadamized.
3 It is reported that the sums kept by the political class and higher bureaucracy can be
  anywhere from 5 percent to 30 percent of the total amount spent.

REFERENCES

Allison, Anne. 2013. *Precarious Japan*. Durham: Duke University Press.
Anand, Nikhil. 2006. Disconnecting experience: Making world class roads in Mumbai.
  Economic and Political Weekly 41(31): 3422–3429.
Appadurai, Arjun. 2013. *The future as cultural fact: Essays on the global condition*. New
  York: Verso.
Appel, Hannah Chadeayne. 2011. Futures, oil and the making of modularity in Equato-
  rial Guinea. Ph.D dissertation, Stanford University, Department of Anthropology.
Barak, On. 2013. *On time: Technology and temporality in Modern Egypt*. Berkeley: Uni-
  versity of California Press.
Barnes, Jessica. 2016. States of maintenance: Power, politics, and Egypt's irrigation infra-
  structure. *Environment and Planning D: Society and Space* 35(1): 146–164.
Baviskar, Amita. 2005. *In the belly of the river: Tribal conflicts over development in the
  Narmada Valley*. New York: Oxford University Press.
Bear, Laura. 2007. *Lines of the nation: Indian railway workers, bureaucracy, and the inti-
  mate historical self*. New York: Columbia University Press.
Bear, Laura and Nayanika Mathur, eds. 2015. Remaking the public good: A new anthro-
  pology of bureaucracy. *The Cambridge Journal of Anthropology* 33(1).

Benjamin, Walter. 1969. *Illuminations: Essays and reflections*. Ed. Hannah Arendt; trans. Harry Zohn. New York: Schocken.

———. 1999. *The Arcades Project*. Trans. Howard Eiland and Kevin McLaughlin. Cambridge, MA: Harvard University Press.

Bhabha, Homi. 2004. *The location of culture*. New York: Routledge.

Borras, Saturnino M., Jr., Philip McMichael, and Ian Scoones, eds. 2010. Special issue: Biofuels, land and agrarian change. *Journal of Peasant Studies* 37(4).

Boyer, Dominic. 2011. Energopolitics and the anthropology of energy. *Anthropology News* 52(5): 5–7.

———. 2014. Energopower: An introduction. *Anthropological Quarterly* 87(2): 309–333.

Carse, Ashley. 2014. *Beyond the big ditch: Politics, ecology, and infrastructure at the Panama Canal*. Cambridge, MA: MIT Press.

Collier, Stephen J. 2011. *Post-Soviet social: Neoliberalism, social modernity, biopolitics*. Princeton, NJ: Princeton University Press.

Ghertner, D. Asher. 2015. *Rule by aesthetics: World-class city making in Delhi*. New York: Oxford University Press.

Gordillo, Gaston. 2014. *Rubble: The afterlife of destruction*. Durham: Duke University Press.

Graham, Stephen, and Colin McFarlane. 2014. *Infrastructural lives: Urban infrastructure in context*. Abingdon, UK: Routledge.

Gupta, Akhil. 2012. *Red tape: Bureaucracy, structural violence, and poverty in India*. Durham: Duke University Press.

Hahn, Hans Peter, Karlheinz Cless, and Jens Soentgen. 2012. *People at the well: Kinds, usages and meanings of water in a global perspective*. Frankfurt: Campus Verlag.

Hansson, Stina. 2010. Responsibilising the state—Methodological considerations of studying the agency of civil servants. Paper delivered to the Studying the Agency of Being Governed workshop, School of Global Studies, Gothenburg University, May 19–20.

Harvey, Penny, and Hannah Knox. 2015. *Roads: An anthropology of infrastructure and expertise*. Ithaca, NY: Cornell University Press.

Hellberg, Sofie. 2010. A methodological framework for a biopolitical reading of local water life hi(stories). Paper delivered to the Studying the Agency of Being Governed workshop, School of Global Studies, Gothenburg University, May 19–20.

———. 2012. Water management and the security-development nexus: The governing of life in the eTheKwini municipality, South Africa. In *The security-development nexus: Peace, conflict and development*, ed. Ramses Amer, Ashok Swain, and Joakim Öjendal, 205–228. New York: Anthem.

Klein, Naomi. 2008. *The shock doctrine: The rise of disaster capitalism*. New York: Picador.

Koselleck, Reinhart. 2004. *Futures past: On the semantics of historical time*. New York: Columbia University Press.

Mains, Daniel. 2012. Blackouts and progress: Privatization, infrastructure, and a developmentalist state in Jimma, Ethiopia. *Cultural Anthropology* 27(1): 3–27.

McMichael, Philip. 2010. Agrofuels in the food regime. *Journal of Peasant Studies* 37(4): 609–629.

Meadows, Donella. 1972. *The limits to growth: A report for the Club of Rome's project on the predicament of mankind*. New York: Universe.

Mitchell, Timothy. 2009. Carbon democracy. *Economy and Society* 38(3): 399–432.

———. 2011. *Carbon democracy: Political power in the age of oil*. New York: Verso.

Öjendal, Joakim, Stina Hansson, and Sofie Hellberg, eds. 2011. *Politics and development in a transboundary watershed: The case of the lower Mekong basin*. New York: Springer.

Rodney, Walter. 1981. *How Europe underdeveloped Africa. Revised Edition*. Washington, DC: Howard University Press.

Schwenkel, Christina. 2013. Post/socialist affect: Ruination and reconstruction of the nation in urban Vietnam. *Cultural Anthropology* 28(2): 252–277.

3

Infrastructures in and out of Time:
The Promise of Roads in Contemporary Peru

PENNY HARVEY

Reconfiguring the Future

In Peru people long for roads in ways that are quite hard to grasp from the perspective of those of us who have always assumed that we can get in a car and drive to where we want to go. Nevertheless their sense of longing (*anhelo*) chimes with the overwhelming sense that infrastructures are the current currency of investment across the planet, the go-to promise for a better future. It is this promise that concerns me here. Working with an understanding of promise as "that which affords a ground of expectation of something to come," I am interested in exploring the grounds of expectation that accompany infrastructural projects. The ethnography on which I draw concerns roads, but along the way I aim to make a more general case for the value of the ethnographic study of infrastructural projects.[1]

The longing for roads in Peru is intimately connected to the perilous state of contemporary provision. Somewhat ironically Peru is renowned for its ancient roads, constructed in the fifteenth century as the backbone of the Inka empire. Long before the appearance of the car, or even the wheel, the Inka were able to eat fresh fish delivered by relays of runners direct from the shores of Lake Titicaca to the imperial capital of Cusco. These roads materialized the administrative axes of empire, enabling the swift movement of armies and consolidating the possibilities of imperial control and integration. However, the roads that concern me here are those sponsored by the modern state, built to integrate a national territory, and more recently to ensure connectivity to wider global markets. These roads are of an altogether different kind from their imperial Inka predecessors, not simply in terms of their (in)capacity to endure over time, but in terms of the temporal horizons that are constitutive of their value. Contemporary infrastructure projects are configured in relation to modern understandings of the future, as a time/space of potentiality for change and improvement.

Reinhart Koselleck ([1979] 2004) has explored the emergence of the modern future in his study of late eighteenth-century European thought as a time when understandings of the relations between past and future were dramatically reconfigured by a whole range of events and inventions—in technological, political, economic, and religious spheres. He argued that modernity was the bringing together of past and future in a new relationship (31) in which past experience did not necessarily dictate future possibility. Indeed the possibility of radical change and the intrinsic unknowability of what was to come produced the foundational modern distinction between the temporal structure of experience and the temporal structure of expectation. The future was open to human intervention, and concepts such as "revolution," "progress," and "development" began to take on new meanings that challenged the certainties of the premodern world, where "the course and calculation of historical events was underwritten by two natural categories of time: the cycle of stars and planets, and the natural succession of rulers and dynasties" (37). The modern future assumed uncertainty and multiplicity.

Koselleck is primarily interested in the implications of this change for the writing of history, for the very purpose and import of the historical account was itself challenged by the disjunction of past and future. The response from contemporary historians was a move away from "history in general" to the assumption of History, as a telos of process and progress. The emergence of History, as a "collective singular," was one of many concepts that came to define the modern. Koselleck refers to the emergence of the collective singular as a philological event: "This philological event occurred in a context of epochal significance: that of the great period of singularization and simplification which was directed socially and politically against a society of estates. Here, Freedom took the place of freedoms, Justice that of rights and servitudes, Progress that of progressions and from the diversity of revolutions, 'The Revolution' emerged" ([1979] 2004: 35).

Thinking of the emergence of the collective singular as a response to the uncertainties and openness of future potentiality resonates with my ethnographic observations of contemporary road-construction projects. Notwithstanding Bruno Latour's convincing claim that "we have never been modern" (1983), the singular collective is alive and well in road-construction projects. Indeed the very notion of "the project" is one of its more ubiquitous forms.

Koselleck's thesis also resonates with contemporary anthropological studies of development practice, where the critique of assumptions about linear connections between the design and realization of a project go hand in hand with the rich ethnographic description of the considerable effort involved in attempts to secure the passage from past to future in acts of design, policy, and planning (Mosse 2004; Simpson 2013). The intrinsic openness of future possibility draws forth the desire to plan and to intervene, but the connection between plans and outcomes is hard-won. The effort of securing a smooth transition from past to future is central to my interest in emergent infrastructural form. Infrastructural projects are invariably both "future oriented" and "future positive" (Li 2007): they promise not only improvement but accelerated improvement afforded by particular modes of connectivity (Simpson 2013). However, the sense of hope and expectation that surrounds these projects never erases the uncertainty, and the promise of infrastructural provision excites both desire and fear of unpredictable and negative consequences.

In what follows I set out to explore the force of infrastructural promise as a complex and unstable temporal alignment, as I look to specify how the relationship between past and future is articulated in and around the practices of road construction. Laura Bear's (2014a) call for an anthropology of time that seeks to go beyond the general and abstract conclusions about time/space compression (Harvey 1989) or the accelerations of modern life (Virilio 1989) is a call for ethnographic attention to the multiple representations, technologies, and social disciplines of abstract time (that is, Time as a collective singular in Koselleck's terms) in relation to both the social and institutional framing of time, and the many and diverse concrete experiences of time.

I am interested here in how roads are conceived and realized as projects, that is, as interventions that take place—and take form—as events, as significant moments in a particular historical trajectory. However, the time of infrastructure has its own plasticity. Projects can take many decades to realize; they pass through many hands, and exist in dynamic tension with the material and social conditions of their own emergence. The singularity of the project is tenuous. What emerges on the ground appears in fits and starts. Some aspects of a project appear long before others, while some components of a system might begin to fail or decompose before others have even begun. It is in this sense that this chapter on road-construction projects in Peru explores infrastructures as situated in time, and yet always in some sense "out of time" with themselves. The time of infrastructural formation folds together many histories and rhythms of material and social transformation, some of which are foreseen, many of which are not. The promise that is harbored by the project also mutates over time, resonating with all those dimensions of life that never were integral to how the project was conceived, articulated, or realized. In short, to understand how roads configure the future in contemporary Peru, it is necessary to attend to the complex conditions of their emergence, to the ways in which specific understandings of the relation between past and future are sedimented in their material form, and to the ways in which these projects articulate expectation and experience in their trajectory toward an emergent future.

Before turning to the detailed description of two specific Peruvian roads, I first elaborate on the analytical purchase of the infrastructure concept, and I ask what difference it makes to approach roads as infrastructures.

The Promise of Infrastructure for Anthropology

The study of infrastructure is booming in contemporary anthropology. New ethnographies of infrastructural formation appear all the time.[2] These accounts are responding to the current ubiquity of infrastructural investment across the planet. New and powerful infrastructural forms mediate the past and the future for many people. The ever-more extensive webs of connectivity appear as tangible forces of transformation that combine technological promise with political will and social uncertainty. The promise of infrastructural transformation is thus ethnographically salient. And even where systems fail to appear, and/or when they break down, an understanding of the potential agency of these systems can nevertheless shape people's everyday material existence (see Appel, this volume).

However, the infrastructure concept is only partially ethnographically motivated. While infrastructure might appear as an ethnographic concept in contexts of either planning or breakdown, the investment of contemporary anthropology in the analysis of infrastructural formation probably has as much to do with its analytic purchase as with its empirical salience. From an analytic perspective, infrastructure is an exemplary relational form, and furthermore a form that mediates diverse theoretical concerns and speaks to contemporary debates on the relationship between human and nonhuman agency, on material politics, event and duration, past and future, space and time. In short infrastructural forms offer a focus for the analysis of contemporary social life that side-steps the limits of humanism without erasing the human, and allows for a dynamic and open sense of scale that does not assume a singular perspective.

Infrastructural systems engage the material turn in anthropology and ethnographies that track infrastructural effects approach infrastructures as part and parcel of the fabric of human life, the systemic conditions of possibility for specific modes of human action. In this understanding of the infrastructural we can assume that any mode or field of human action relies on infrastructural arrangements that are not the subject of direct reflection within that field of action. Thus the "infra" quality of an infrastructure does not necessarily refer to invisibility. Roads are there for all to see. It is rather that the systemic form of an infrastructure is both self-sustaining and backgrounded. A road is infrastructural to a driver, but not to those charged with maintaining the road surface. In practice, the origins of infrastructural forms will influence their relative visibility, because actions and motives are rarely singular. Infrastructural forms might be deliberately produced as acts of spectacular engineering (e.g., Dalakoglou 2012; Carse 2014), openly combining a material and an aesthetic function, designed to enthrall as well as to connect. They are human artifacts with origins in deliberate attempts to create a systemic relational form. But as infrastructures (rather than as monuments to specific acts of human ingenuity or stupidity) they do not need to be visible. If the infrastructure works as infrastructure, it can carry out its relational work in a less dramatic way. In many cases it is indeed preferable for the system to sink underground, to work behind walls, or to connect via vibrations or electromagnetic forces. The point is that an infrastructural system should not require ongoing attention from agencies external to its constitutive relations.

As others have noted it is when infrastructures stop working that they appear as systems that need more than their own integral relational dynamics to keep them going. On the whole in the infrastructure literature, any "breakdown" is approached as the limit point of infrastructural flow and is exceptional (however common!). Where distributed complex systems fail there are rarely singular causes, and there is much to learn from the archaeology of failure.[3] However in this chapter I attend to disjunctures of a different kind. These are the disjunctures internal to the working system. Drawing on the ethnographic research that Hannah Knox and I carried out regarding road construction in Peru, I present here an approach to infrastructural formation that privileges attention to internal differentiation and disjunction, a position that emerges from the general acknowledgment within contemporary anthropology and science and technology studies (STS) that infrastructures are best approached as assemblages (unstable and noncoherent gatherings of heterogeneous materials, skills, and practices). My interest in exploring this possibility emerges both from the analytical challenge of articulating the specificity of the social dynamics of particular assemblages (assuming the assemblage as a starting point rather than as a conclusion), and the analytic demands thrown up by our ethnographic encounters with engineering practices.

Engineers do not in my experience deploy the language of noncoherence, but they are attentive to the interruptions and interferences in the systems that they create. These interruptions and interferences derive from the dynamic and complex multiplicities internal to any specific system, and from the self-evident fact that no system exists in isolation. As we consider the recursive quality of infrastructural relations—the enfoldings and overlaps of multiple coexisting infrastructural systems—what comes to the fore is that there is no ultimate technical solution that would allow an engineer to produce a self-sustaining, smooth system. Elsewhere we are quite familiar with the notion of the needs for upgrades in software systems, the fixing of bugs, and the unfolding changes in specified functionality. My point is that even when couched in the language of the technical (and engineers are always willing to address the technical) the "problem" is social and derives from the intrinsic connectivity of open-ended structures.[4] It is the relational capacity of infrastructural form that both sustains and undermines the expectation of a self-sustaining system, a technical fix, or the promise of total automation. Indeed the key task for engineers is often to find a way to contain and limit the multiplicity, the dynamism and the expansiveness of the systems they create (Harvey and Knox 2015a). In short, and to return to the question of the difference it makes to approach roads as infrastructures, if the road-construction project exemplifies the collective singular of development paradigms, the infrastructural *promise* exemplifies a relational social form that is both systemic and internally differentiated, both coherent and emergent. I turn now to the two roads in Peru that drew me into thinking about the power of the infrastructural analytic.

The Promise of a New Road

I first went to the small town Ocongate in 1983. The road at that time was in a terrible state. Heavy trucks traveled the seven-hundred-kilometer route from Cusco in the Andean highlands to the Amazonian town of Puerto Maldonado. In many places the road was too narrow for the trucks to pass. The direction of traffic was supposed to alternate day by day, but breakdowns, landslides, and accidents were commonplace and there were always counterflows and hair-raising passing points requiring one vehicle to inch out to the edge of a precipitous drop. Many died on the roads. The dangers were considerable, especially for those without the option to choose a sturdy, well-maintained vehicle or a reliable driver. Despite the dangers there were always many passengers perched on top of the cargo.

So, the road was bad, but its history was fundamentally concerned with trade and movement, and in that sense we could retrospectively and analytically approach this road as an infrastructural form. From the beginning it stimulated and supported international circulations not only of products but also of people looking for opportunities to make money. I did not initially pay much attention to the road beyond registering both the fear and the enjoyment of protracted and uncertain travel. My focus on roads emerged as an integral part of a wider interest in the modern state, and the challenge that its distributed and ephemeral presence posed for ethnographic methods. The material condition of the roads, their very presence as well as their state of repair, registered histories of both state presence and neglect. For Hannah and me, the roads appeared as barometers of the lived experience of global political economy in the ways their material fabric registered the dynamic movements and intersections of people, goods, and markets, and the fluctuating political will to build surfaces that could resist the penetration of water and frost. From the perspective of state planning, the histories of road construction displayed the oscillating and contradictory impulses of territorial integration and deterritorializing extension. From the perspective of local people, the road registered gestures of state care and a sense of inclusion. However, when the surfaces broke up, when overloaded vehicles carved deep trenches in the mud that rendered the route impassable, and when people died because they needed to keep going anyway, then the very same roads provoked an overwhelming sense of abandonment and isolation.

The roads were rarely associated with moments of decisive state action. Long stretches of the existing road between Cusco and Puerto Maldonado had been constructed either by local communities or by local landlords without any external support. In the Andean region the forces of the Earth and of the mountains, living beings fundamental to the productive capacity of human enterprises, were also drawn in. Road construction entangles the force of human labor, the supernatural interventions of Catholic deities, and the other-than-human energies of the Earth and in the process it involves a series of intersecting temporalities that go well beyond the narrative of engineering design or political calculation. Roads typically emerge over time in a punctuated process, and narratives of this emergence connect the erratic and unpredictable energy of state agency to the equally uncertain powers and demands of a sentient landscape.

In 2006, over twenty years since I first arrived in Ocongate, the road project had just reappeared in a new and compelling guise. There were plans to build a major international highway connecting Cusco to the frontier with Brazil (at Iñapari, some two hundred kilometers beyond Puerto Maldonado). The new road would more or less follow the existing route, and Ocongate was quite clearly going to be affected. As yet though, things remained much as they had ever been. A protracted civil war had precluded state infrastructural investment for many years. Now there was political will to support development infrastructures for economic revival and political integration. Journey times had shortened somewhat over the previous decade and old buses were in use, allowing people to travel in a bit more comfort. The new road seemed set to deliver time-space compression of an entirely different order. A fast surface and a widened platform would transform Ocongate into a settlement only two hours away from the vibrant tourist economy of Cusco.

The new road was enthusiastically supported by many different constituencies. The Brazilian state had already constructed a fast road up to the Peruvian border where it awaited the link to newly built mega-ports on the Pacific coast through which they intended to export soy to China and the growing Asian markets. At the national and regional levels there were many discussions about extending Cusco's tourist trade into this region, and expectations (and concerns) were raised about increased circulation of products to and from local markets. Locally the desire was simply for a safer, faster road. Nobody could deny the possibility of the new interconnections that the road would make possible, the desire that life might become a little easier, a little less precarious, a little less exhausting. The new road offered the possibility to build new futures of many different kinds simultaneously. Indeed, the decision by the Ministry of Transport to finance the loan for the building of this road was taken in the knowledge that the project was widely supported by key constituencies, at international, national, regional, and local levels. Critics far away in Lima argued that this was primarily a political road, strategically deployed to create an effect of political consensus at a time of chronic instability. The economic rationale was never clearly articulated, and there were fears that the government was wasting money on a region that, while lacking in investment, did not stand to gain sufficiently from this initiative to make it worthwhile. This argument cut no ice at the local level. The promise held strong despite the lack of any systematic analysis of the potential benefits. Furthermore, guided by the construction company, the project took shape around a particular temporal paradigm of development and progress. The promise was of forward movement, of accelerated transition from the past to the future. Images of before and after were displayed on billboards, on websites, and in PowerPoint presentations at all public gatherings. Sleek new trucks and even sports cars were depicted. The future was to be both shiny and fast moving.

The Prospect of a New Road

The other road that Hannah and I studied did not appear at first to be taking people into the future at any speed. The road, which lies in the northern Peruvian Amazon and runs for one hundred kilometers between the city of

Iquitos and the town of Nauta, intrigued us because it did not connect to any other roads, and its narrative was not overtly one of international connectivity. Its potential lay in the links it provided between the navigable rivers of the Amazon basin and the international airport at Iquitos. However when we first encountered this road its infrastructural status was somewhat dubious, as its connectivity was not as yet established, and its future seemed quite uncertain. The road had been embroiled in a disproportionate degree of scandal and corruption over the years and there were still ten kilometers of highly unstable surface.

From our discussions with the small construction company charged with the completion of the road we assumed these final ten kilometers would soon be finished. But all those we spoke to both locally and in Lima insisted that nothing was likely to happen anytime soon. There was ongoing litigation involving the regional government and various construction companies. This short road had been "under construction" for the past seventy years and the general opinion seemed to be that it could be seventy more before any smooth infrastructural connection between Nauta and Iquitos could be counted on. In 2005 there was some work going on, but the unfinished parts of the route were so muddy that vehicles could only get through if dragged by the heavy machinery of the construction company. Another problem was that in places the surface of the road that had supposedly already been completed was of such poor quality that it had already begun to crumble. Corruption accusations had led to the impounding of all materials and machines belonging to a former construction consortium, and we were told that the delay on the final stretch was because there were five kilometers where no construction could take place until all the material evidence had been gathered and the lawsuit settled. Further delays had been caused by a major dispute over labor contracts that had led the residents of Nauta to storm the construction camp and demand the removal of the engineer heading up the construction team.

The optimism of the construction company that we were talking to seemed to reflect the fact that they had only recently been contracted. They were working steadily toward completion of all but the disputed five kilometers, but they had their own troubles. Heavy rains and strong winds were tearing up the newly laid surfaces and they were still testing out new materials (with enhanced stabilizing properties) and building improvised shelters to protect the new surface from the inclement conditions. In short, the construction process seemed to elicit endless material complications, legal disputes, and social disturbances. The more we found out about this road the more we came to understand the view that nothing was likely to be completed any time soon.

But suddenly, and while we were away, the road was finished, and before we knew it an inauguration ceremony had taken place, involving the president of the republic, Alejandro Toledo. After the seventy-year hiatus we were somewhat miffed to miss this final chapter. But we were not the only ones. Emilio Puma also missed the ceremony. Emilio had been introduced to us when we had started expressing an interest in the road. He was one of the pioneers—a group of men, now in their eighties, who had been the first to trace the route through the dense forest at a time when all movement between Iquitos and Nauta had been by river. Officials from the regional government had told Emilio that he would be an honored guest at the inauguration ceremony and that he would be presented to President Toledo. This was exciting news. Over the years the pioneers had seen the road grow from an extremely local project to become a major public work that had brought international finance capital, engineering expertise, and political controversy to the small town of Nauta. The pioneers were proud of the crucial role that they had played in the initiation of this project. It was a local politician, the mayor of Nauta, who had originally contracted the pioneers to find the route. In those days the road out of Iquitos only ran for twenty-one kilometers. That section had been constructed in the 1930s, but since that initial foray to extract hardwoods from the forest there had been no attempts to extend the road to Nauta. The residents of this small town spoke of their seventy-year longing for the road. This longing also dates back to the 1930s when the Amazon River moved—as rivers do—stranding what was then a thriving river port on the edge of a sandbank, its modernizing ambition thwarted, while the town of Iquitos took their trade and grew into a large city. The pioneers are thus key figures in the story of how the residents of Nauta struggled to reinstate their connection to the modern world. Emilio was at pains to point out that all the initiative for this project had come from the people of Nauta: "Things get done around here *because* there is no support from the authorities. They don't help you, so you learn to do things for yourself." From the perspective of the pioneers, the road had emerged thanks to local commitment and energy. But the pioneers did not feel recognized. They were never paid for the work they did, and there was some resentment that despite their fundamental contribution to what had subsequently become a project of national interest, they had received neither pension nor recognition. The inaugural event seemed as if it might put some closure on this long-term waiting—devoid of any immediate expectation—and usher in a new set of possibilities for themselves and for their town.

Unfortunately things did not turn out well. President Toledo was due to arrive at kilometer 5, and the pioneers had been instructed to wait in Nauta for somebody to come and collect them but nobody came. When they finally got to the ceremony they were not allowed through the security cordons and by the time that word had got through to those who could have afforded them some visibility in the proceedings, Toledo was no longer there. He had arrived and left in a helicopter as presidents do. His visit had been brief. He had smiled for the cameras, made a short speech about the commitment of his government to the development and progress of the region, and he had moved on to his next appointment. The gap between the pioneers and their president gaped before them. Even when in the vicinity, drawn there by the precious road in which so much hope had been invested, the direct connection between local aspiration and central state preoccupations had blatantly failed to materialize.

The promise of the road and the promise of some recognition and concern from the state are entangled in this story, raising the question of what exactly infrastructural connectivity is expected to deliver. In both examples, and in ways that I develop further in the final section of this chapter, there is an important difference between the delivery of the *project*—a specified, material connection between nodes (place A and place B)—and a general and under-specified notion of enhanced connectivity to a more productive and secure life world, the "modern" world of the future positive.

The Time of Infrastructure

These two brief accounts give some sense of how our study approached the roads of Peru as diagnostic spaces that could offer us an empirically grounded perspective on how the fluctuations in state policy, regulation and practice, corporate finance and engineering expertise, and the expectations and efforts of local people all coexist in infrastructural systems. We treated the roads as complex sites of political intent—emerging over many decades of irregular investment and changing priorities and ambitions. Needless to say, there were also crucial and unforeseeable processes that became integral to the roads as they exist today. Nobody planned the gold rush of the southern Peruvian Amazon, or the illegal timber trade. There was no modeling on the movement of drugs, the resettlement of forest communities, the speculation in land, the move away from the rivers. Roads channel and constrain circulation, but as infrastructural systems they do so in ways that are open-ended and subject to continual modification. Infrastructural systems are intrinsically relational, and their ever-emergent form and their dynamic social value imply the assemblage of multiple differences within the materials, institutions, regulations, aspirations, and skills through which they are constituted.

Complex temporality is integral to the infrastructural assemblage. This relationship between the "what" and the "when" of infrastructural systems was the subject of an influential article published in 1996 by Susan Leigh Star and Karen Ruhleder. Aware of the complexity and the multiplicity of infrastructural systems they were keen to emphasize that the *what* of an infrastructure depends on the social and material relations through which it is put to use. Infrastructures—like any other tool—are always operationalized in specific practices. The question "*when* is an infrastructure?" thus directs the researcher to look at the diverse and particular taskscapes that technical systems facilitate and, equally, the conventions and standards that both support its systemic quality and limit the uses to which it is put. Star and Ruhleder's focus on the when *also* points to the ways in which all kinds of material and social arrangements could become "infrastructural," but equally might fail to do so. They identified the specificity of the infrastructural as the capacity to operate seamlessly across scales (spatial, temporal, material, conceptual) simultaneously supporting extensive and local ventures.

The roads we studied in Peru were built with the ambition to integrate and thus reconfigure relations at multiple scales. In the process they produced experiential realities that went beyond that which could be concretely assumed before the roads were built. Linking Iquitos to Nauta re-created the kinds of places that Iquitos and Nauta were and also reconfigured their potential future. Similarly the smooth paved road, on which Ocongate now lies, affects present experience and shapes future horizons.

Star noted that the intrinsic multiplicity of infrastructural systems generates what she referred to as the "trouble in the system." That trouble points to internal tensions that, strictly speaking, defer the when of the infrastructural. But these disjunctive temporalities are partial disconnections to which a system can adapt. The costs of such adaptation, however, often imply the erasure of that which might otherwise have been included. For example, a road built to support high-speed international trade will not necessarily be compatible with a road that is expected to foster local development—as the residents of small towns along the interoceanic highway have found out to their detriment.[5] In Ocongate a road that was brought as close to the town center as was possible without destroying the very fabric of the town later became a bypass. Ocongate is now too near to Cusco to be the optimal place for a traveler to rest, so the possibilities for trade have shifted in ways that people are still coming to terms with.

Civil engineers that I have worked with in both Peru and the United Kingdom have stressed to me over the years that engineering and construction are two quite different things. "Engineering" is the production of the technical solution, design, or prototype that is subsequently realized in the practice of construction or manufacture. In this respect, the crucial engineering work is carried out prior to the construction phase of a project. A feasibility study for projects such as those we studied in Peru will include all kinds of technical details and specifications concerning the road surface, its foundations, its materials, its form, its routing, and its potential costs and benefits—including the calculated risks with respect to environmental and social impact. Technical studies combine the calculative and the speculative in a specific mode of anticipation. The road will be designed in relation to a projected life span and a projected life condition (the numbers of vehicles, their speed and weight, the repeated and cumulative stress on the surface, the effects of seasonal and diurnal cycles, of heat and cold, of water and drought). These predictive processes are abstract and built on statistical projection. Modeling possibilities is, of course, quite different from knowing what is going to happen. Engineering design nevertheless offers a point of stability in the construction process. Despite the volatility of some of the technical data, to which I return shortly, the key stabilization that these studies achieve refers to the clarification of the agreement between funders, politicians (and/or representatives of diverse public constituencies), and the construction company. The engineering design will address questions such as "what kind of a road do we want and expect this to be?" and "will this project be related to other concerns and policies with respect to public investment?" The studies are ultimately consolidated in the contractual agreements that are drawn up to allow a project to start through the release of funds.

Quite different knowledges are assembled in the laboratories of the construction companies once the works are under way. Here the key relations to set out and sort out are those between the available materials and the environmental forces that they will be expected to confront. The engineers measure and model the relational capacities of the materials with respect to things like their relative resistance to weight, plasticity, or porosity; they gauge the relative value of natural over man-made materials; and they work to find the best fit between what they have to work with and the agreed specification of their final product. These are still activities that move toward the production of an engineering design. The temporality has shifted, however. The design is still prior to the subsequent act of construction, but it is also recursive (Kelty 2005). The problems and challenges that the construction process produces are continually referred back to the laboratory for modification and subsequent design refinement.

The notion that the engineering work is always prior to and separate from construction is thus itself a somewhat abstract or ideal account of how construction proceeds in practice. Engineering design informs the construction process, and it signals what it is that has to be built and how, but the design remains open to modification. The technical studies direct proceedings, but they do not in the end determine how to proceed. The realization of the design is something that is worked out on the ground in the interactions between engineers (contractors and supervisors), foremen, laborers, materials, and machines. The importance of the technical specifications lies in their capacity to clear the way for action by setting out the parameters of the material transformations that are to be undertaken. They also serve to delimit the relational domain for which the engineers are responsible. In this respect we see that the framings entailed in the drawing up of a technical specification are not predicated on the failure to address local conditions, but could perhaps be seen as attempts precisely to localize the space of intervention, to articulate its specificity and to limit responsibility for all that will, inevitably, overflow this space at some unspecified future date. In this respect technical knowledge in the design phase works by closing down alternatives.

In many respects the construction phase unravels the certainties of design and then reverses this sense of partial closure. We also find different temporal regimes taking over from the speculative calculations and statistical projections. On the one hand there is considerable effort put into the logistics of the project. Logistics are the systems deployed by the construction companies to ensure that the works stay on schedule and within budget. Logistical work is the work of coordination, categorization, and scheduling. In the continual adjustments that are made as to who will do what, when, and where, time assumes a monetary value. The logistics configure a matrix of value where time is money. The complex internal agendas that a company manages does, however, allow that not all decisions take the most direct route to profit. Good relations can bring their own advantages and the construction company that allowed us to hang around as resident anthropologists also quite frequently accommodated local demands that were not strictly within the terms of the contract. They also at times reacted strongly—even violently—to persistent demands or behaviors that were deemed to be prejudicial to their progress.

In addition to the logistical mode of temporal ordering, the other important temporal dimension of the construction process was the anticipation of the unforeseen. Construction engineers know and anticipate that the designs will not in practice guide them smoothly from A to B. They know that getting from the present to the future is not simply about good planning or efficient logistics. Quite apart from the volatility of the social environments through which these road-construction projects progress, often over several decades, the material environments offer challenges that the engineers only really solve once they are involved in the transformative process. Rivers move, hillsides collapse, and rains, frosts, and winds impede progress. Unlike the designers working on computers in remote offices, the field engineers work in a more direct relation to that which they seek to transform. In this respect there is a strong awareness that some things take time. Construction becomes more of a craft practice at this stage in the accommodations to the materials as well as to the diverse human demands and practices that bring about delays and disruptions.

Inaugural Events

This tension between the public visibility of contemporary infrastructural systems (as promising or threatening technological configurations) and the mundane assumption of the "infra" status of the smoothly functioning space of flows, which passes unnoticed for those whose activities it supports without disruption, brings me back to the discussion of inauguration ceremonies for public works.

Both the roads we studied in Peru were visited by presidents who sensed the electoral (or legacy) advantage of a close association with these promissory spaces. Indeed, on the interoceanic highway there were even reports of disputed paternity as successive presidents staked their competing claims to having brought the road into being. President Toledo opened the first kilometer, which was completed and inaugurated before the rest of the road even had a full engineering profile. The urge of presidents to inaugurate public works indicates the political purchase of infrastructural projects. It also indicates that what is celebrated is not social transformation per se. The inauguration events certainly anticipate a positive transformation, but state actors are tentative about overidentification with specific future outcomes. Inaugural speeches tend to focus more directly on the resources that have been brought to an area. The metrics of investment index state commitment. The political rhetoric focuses on figures that tell us how much was spent, the kilometers covered, the number of bridges built, or the numbers of jobs and of potential beneficiaries. But these metrics are rehearsed as metrics of care and political investment in unspecified futures.

Inaugurations are public rituals that seek to fold construction work, which an engineering company and often a fairly large and dispersed workforce have brought about, back into a narrative of state-led development. In these rituals the infrastructure is staged as an event. The inauguration of the Iquitos-to-Nauta road, discussed earlier, ostensibly celebrated the completion of the project. The work had taken form and created a smooth connection between two places, responding to the seventy-year longing on the part of the people of Nauta. But as we have seen, the scalar complexities internal to this project were still troubling. The road was visibly unsettled, far too contentious and discontinuous to assume the status of an infrastructural system that had transcended its scalar discontinuities. It was certainly quite unclear how this connection between Iquitos and Nauta might come to underpin national, regional, or local futures. The promise of the road hinged on its transformational potential, but the ongoing material disintegration and the lawsuits and the many social conflicts that the new road generated suggested that the inevitable "trouble in the system" was such that its future effects were very uncertain. Even the president who *could* have driven to the site of the inauguration had chosen instead to travel by helicopter. In short this was not a road that was in need of repair due to the negative effects of external forces. It was a road that for many had yet to assume its infrastructural form. The sense of incompletion—and non-event—was integrally linked to its expansive and diffuse promise. The surface was visible but its infrastructural force remained obscure.

Inaugural events carry a very particular kind of promise—a promise that extends the notion of the compound presence of infrastructures to include spatial and temporal disjuncture. The promise of public works in Peru is routine in the electoral repertoires of politicians at all levels, and from all points in the political spectrum—and conflates the promise of a technology with the promise to deliver a public good. The inauguration is of a different order. The augury, as Frédéric Keck and Andrew Lakoff point out (2013), is a "figure of warning." The inauguration ceremony attempts to render the new beginning auspicious. Standing at kilometer 5 on the Iquitos-to-Nauta road, Toledo had endorsed the new road as testimony to his presidential status and, as Max Gluckman (1940) led us to expect, the inaugural ritual reproduced the finely differentiated social space that distinguished those who were there from those who were not, and it shaped how they were there, how they arrived, and how they left. But the disappointment at the heart of Emilio's story also brings home how the inauguration of these infrastructural spaces can launch new futures that are from the outset highly problematic—even inauspicious—in local terms. If the president has come to open a road, he has, by definition, come in the name of a national or even an international project. The ceremonies endorse a scale in which local interest is irrelevant, superseded or transcended by the force of a "public work" that may be located but is no longer local. The quality of the future thus hangs in the air—and, I would suggest that this hanging, this wish that the project might be successful, also makes evident that it might fail. The *promise* of an infrastructural provision can be used to hold a politician to account—if the material form is not provided, support can be withdrawn. But the *inauguration* registers the uncertainty, acknowledging that something more than the form is needed for the infrastructure to deliver that which is expected of it. Its relational potential has to be activated.

Conclusions: Infrastructure as Non-event

It is not my intention to suggest that change does not occur. The issue is rather that the project to transform the material conditions of people's lives will have infrastructural consequences that designers and planners can never control. The eventual quality of infrastructural form remains forever in question, because a project holds together multiple infrastructural possibilities, and the event of transformation can thus only be apprehended in retrospect. In Ocongate, transformations began long before construction started. Local businesses began to gear up for the new trade and the new trading conditions that the arrival of a huge transnational construction project implied. At the same time, local people began to campaign over routing and employment in order to maximize the possible economic benefits. Land speculation was rife and squatter settlements appeared at strategic sites. Rumors circulated about both the operation of hidden interests and the production of uneven advantage. There were also concerns about the specific nature of material transformation—the exact materials that would be needed, the siting of the quarries and the dumps, the potential effects on water courses and the implications of the gashes and cuts to the Earth, the possible increase in landslides and floods. Questions about what exactly people wanted to transform and what they wanted to hold onto multiplied as the works proceeded. These concerns revealed the complex multiplicity of the relational promise. As these expectations began to bear down on the construction company, the engineers found themselves forced back on the exact terms of the contract, the specifics of the project, the commitment to deliver a road surface of a predefined specification. It was not in the "public interest" for them to be diverted by the exponential growth of public demands, because every attempt to singularize a possibility led to further possibilities as this final example illustrates.

I had often gone with my friend Francisca to graze the cattle. She told me that the construction company was building a camp to house the workers down by one of her fields so we decided to take the cows down there by way of having a look at what was going on. The field on one side was bordered by a river. As would be usual at that time of year (it was June), the riverbed was almost dry. A steady stream of trucks was loading up the stone—and carrying it out to a flat piece of land designated as a storage site where stones were sorted by size, as the materials were sifted through giant sieve-like structures, categorizing the aggregates ready for the production of concrete. On the other side of the field the perimeter fence of the new camp was already built, and inside the installation of the prefabricated housing units was under way. Francisca knew all about these camps, although to my knowledge she had never visited one. Others she knew had worked on road-construction projects before. She explained that there would be dormitories and dining areas, toilets, showers and kitchens, places for machinery. She wondered where exactly they would house the prostitutes. Quite undeterred by my argument that prostitutes would have no formally designated space within the camp and would be more likely to work informally outside the camp perimeter, she made it clear that there were many things I would not know, including the names of those making a profit from the lease of the land, or those standing to gain from the rental of their trucks. But there were also many things that she wondered about. What would people eat? Who would get the work? Where exactly would the road go—through the middle of the town, round the edge, down the side valley (where she lived), up on the high plateau above the town? And then there was the question of whether the river would recover. Would the removal of the stone increase the rate of erosion? Could she make a claim on the company for loss of land where the trucks had been working in the river? Would the Ausangate mountain be offended by the dredging of the river whose waters carried vital energies from the mountain to the farmlands? What would the Earth make of the gouging and the cutting that road construction implied? We mused about these things slowly, on and off over the course of the day, not as issues of great concern, just as things to think about. On our way home she pointed to a pool of water that had gathered in a space formed by the extraction of stones from the river. She told me that her brother-in-law was going to ask the company to dig it out a bit further, so that he could farm fish there. It would be easy for them. All the machinery was there.

Francisca's casual contemplation of the infrastructural possibilities that the road-construction process afforded the people of Ocongate had little to do with the actual "project." The construction company struggled to contain the demands that the works threw out as possibilities for material transformations that could serve as new infrastructural possibilities. These new possibilities were created by the road-construction process but were not otherwise linked to the road project. Francisca's idea about the fish farm occurred to her as she observed the camp and the dredging; it was not an idea formulated as a project. It was simply a way of thinking about the ongoing process of transformation through which the material conditions for life in that place take form. The road would by definition become infrastructural to people's lives but without thereby implying any sense of either progress or direction. In this example, the road-construction process was not addressed as an event, a notable break or moment that heralded a new beginning. It was rather apprehended as an ongoing unfolding of potentiality, a non-event that may or may not take form in the future.

Construction engineers also recognized the mundane process of unfolding infrastructural relations. An engineer once explained to us that "a road is like a person—it is not static, it is dynamic—it grows. Every day you learn more about its problems." However grand the project in terms of design and control, and however invested a particular expert might be in specific modes of calculative reason, these road-construction engineers displayed a pragmatic orientation to the worlds that they engaged. They knew that all material entities constantly differentiate and transform, and they understood that the key question to ask of any space of intervention is not what something is, "but what it is turning into, or might be capable of turning into" (Jensen and Rødje 2010: 1). Engineers are attuned to the "when" of the infrastructural. However, this awareness involves engagement in several divergent "temporal regimes": the anticipatory logics of statistical projects and logistical planning; the engagement with a *project*, a linear notion of material and social transformation; and the recognition that time is also the enduring condition of material things. This recognition of duration requires the engineer to also acknowledge a mode of waiting that runs counter to the anticipation and urgency of project management. Thus even when contingency plans and risk-management plans are in place (as they will be, given the preoccupations of those concerned with logistics), anticipation of the unexpected encourages the field engineer to exercise a nondirectional (unplanned) curiosity with respect to the world as it is encountered.

Local people, such as the pioneers from Nauta or my friend Francisca, combine anticipation and endurance in different ways. Local people are often engaged in explicit political struggles to bring a road into being. Purposeful campaigns are projects that also require logistics and planning, calculated and coordinated action, attitudes that are not assumed only by modern engineering companies. They also assume that enduring and waiting without expectation is part of the process. In Nauta, as in Ocongate, the longing for an as-yet-undefined mode of participation or positive inclusion in a wider regional, national, or international modernity was the backdrop to daily lives where people basically got on with other things.

I have suggested that the singularity of any infrastructural project is always undone by its unruly and expansive promise. The promise of infrastructural projects rests on the open possibility of that which has yet to happen. At the start, the project and the promise can seem to come together and to articulate divergent concerns and expectations. The promise allows projects to get off the ground, giving body and energy to the notion of the collective singular, as the means of addressing uncertain futures. However, my ethnographic reflections on the inaugural ceremonies, staged as events to mark the time of infrastructural formation, also bring the non-event of infrastructure more clearly into view. In both the cases described here, the "trouble in the system" remains unresolved. Indeed, I have argued that the when of infrastructural form will always also imply a deferral, a further waiting, a renewed or even a crushed expectation. As others in this volume have explored, infrastructures gesture toward uncertain futures, even as they attempt to stabilize and channel the grounds of possibility for future lives. It is in this space of promise, integral to the project yet always exceeding its confinement that infrastructures provide us with sites from which to trace the relational dynamics and the political processes through which these futures are brought into being.

NOTES

1 This chapter draws extensively on the ethnography of the roads of Peru that I researched and wrote together with Hannah Knox (Harvey and Knox 2015a). The arguments on time that I put forward here were not elaborated in detail in the book. However, all the ethnography and much of the analytical work on infrastructural promise have been worked out in collaboration with Knox.

2 I cannot list them all here! Some useful overviews include Larkin (2013) and the following collections: Harvey and Dalakoglou (2012), Lockrem and Lugo (2012), Campbell and Hetherington (2014). In addition to the current volume, Harvey, Jensen, and Morita (2017) offer an extensive collection that combines anthropological and STS approaches to infrastructural formations.

3 I am thinking in particular here of some of the fascinating literatures that have emerged after the global financial crisis, such as the work convened by Karel Williams on remaking capitalism (http://www.cresc.ac.uk/our-research/remaking-capitalism/). But we could also turn to more established accounts from within STS and anthropology such as Latour (1996) and Laura Bear's recent thoughts on "accidents" (2014b).

4 Recent works on the exponential growth of smart technologies—the relaying of sensory information through digital means—and big data show that the intrinsic

problems that engineers face are not how to enable connectivity but how to produce meaningful (value-added) connectivity. See Ruppert et al. (2015).

5 See Thévenot (2002) for a discussion of the incompatibilities internal to a road-development project in the French Pyrenees.

REFERENCES

Anand, N. 2011. Pressure: The PoliTechnics of water supply in Mumbai. *Cultural Anthropology* 26(4): 542–564.

Bear, L. 2014a. Introduction to *Doubt, conflict, mediation: The anthropology of modern time. JRAI* 20: 3–30.

———. 2014b. For labour: Ajeet's accident and the ethics of technological fixes in time. *JRAI* 1(20): 71–88.

Campbell, J., and K. Hetherington, eds. 2014. Nature, infrastructure and the state in Latin America. Special issue of the *Journal of Latin American and Caribbean Anthropology* 19(2).

Carse, A. 2014. *Beyond the big ditch: Politics, ecology, and infrastructure at the Panama Canal.* Cambridge, MA: MIT Press.

Dalakoglou, D. 2012. "The road from capitalism to capitalism": Infrastructures of (post) socialism in Albania. *Mobilities* 7(4): 571–586.

Gluckman, M. (1940) 2002. "The bridge": Analysis of a social situation in Zululand. In *The anthropology of politics: A reader in ethnography, theory, critique*, ed. J. Vincent, 53–58. Oxford: Wiley Blackwell.

Harvey, David. 1989. The condition of postmodernity. Oxford: Blackwell.

Harvey, P., and D. Dalakoglou. 2012. Roads and Anthropology. Special issue of *Mobilities* 7(4).

Harvey, P., C. B. Jensen, and A. Morita, 2016. *Infrastructure and social complexity: A Routledge companion.* London: Routledge.

Harvey, P., and H. Knox. 2015a. *Roads: An anthropology of infrastructure and expertise.* Ithaca, NY: Cornell University Press.

———. 2015b. Virtuous detachments in engineering practice—on the ethics of (not) making a difference. In *Detachment: Essays on the limits of relational thinking*, ed. M. Candea, J. Cook, C. Trundle, and T. Yarrow, 58–78. Manchester, UK: Manchester University Press.

Jensen, C. B., and J. Rødje, eds. 2010. *Deleuzian intersections: Science, technology and anthropology.* Oxford: Berghahn.

Keck, F., and A. Lakoff. 2013. Figures of warning. In issue 3: Sentinel Devices. *Limn.* http://limn.it/figures-of-warning/.

Kelty, C. 2005. Geeks, social imaginaries, and recursive publics. *Cultural Anthropology* 20(2): 185–214.

Koselleck, R. (1979) 2004. *Futures past: On the semantics of historical time*, trans. K. Tribe. New York: Columbia University Press.

Larkin, B. 2013. The politics and poetics of infrastructure. *Annual Review of Anthropology* 42: 327–343.

Latour, B. 1993. *We have never been modern*. New York: Prentice Hall.

———. 1996. *Aramis: Or the love of technology*. Cambridge, MA: Harvard University Press.

Li, T. M. 2007. *The will to improve: Governmentality, development, and the practice of politics*. Durham: Duke University Press.

Lockrem, J., and A. Lugo, eds. 2012. Infrastructure. A *Cultural Anthropology* curated collection. https://culanth.org/curated_collections/11-infrastructure.

Mosse, D. 2004. Is good policy unimplementable? Reflections on the ethnography of aid policy and practice. *Development and Change* 35(4): 639–671.

Ruppert, E., P. Harvey, C. Lury, A. Mackenzie, R. McNally, S. A. Baker, Y. Kallianos, and C. Lewis. 2015. Socialising big data: From concept to practice. CRESC Working Paper 138. http://research.gold.ac.uk/id/eprint/11614.

Simpson, E. 2013. *The political biography of an earthquake: Aftermath and amnesia in Gujarat, India*. London: Hurst and Company.

Star, S. L., and K. Ruhleder. 1996. Steps toward an ecology of infrastructure: Design and access for large information spaces. *Information Systems Research: A Journal of the Institute of Management Sciences* 7(1): 111–134.

Thévenot, L. 2002. Which road to follow: The moral complexity of an "equipped humanity." In *Complexities: Social studies of knowledge practices*, ed. J. Law and A. Mol, 53–87. Durham: Duke University Press.

Virilio, P. 1989. *War and cinema: The logistics of perception*. London: Verso.

4

The Current Never Stops:
Intimacies of Energy Infrastructure in Vietnam

CHRISTINA SCHWENKEL

Resilient Infrastructures

In the 1970 documentary *Dòng điện không bao giờ tắt* (The current never stops), the director Đào Lê Bình provides rare footage of the deliberate targeting of urban infrastructure in the Democratic Republic of Vietnam (DRV, or North Vietnam) by U.S. warplanes in an effort to disrupt the country's internal system of supply and distribution. The setting is the Soviet-built electric plant in Vinh City, the capital of the province of Nghệ An in the southern part of the DRV. The film begins by panning the industrial landscape of burning ruins and smoking craters. The plant still stands, but workers, mostly women, are running around, desperately trying to extinguish the fires.[1] Scenes of the chaotic destruction are followed by footage of a group of men shooting at attacking bombers, one of which plummets to the ground. As the planes approach and the alarms sound, plant employees descend quickly into trenches and assume their positions at anti-aircraft artillery, while technicians continue to labor frenetically to repair the turbines and continue the generation of energy. The scene then cuts to another work unit straining to move large pieces of the plant's machinery to a remote forest location as part of a comprehensive strategy of infrastructure evacuation and relocation. Both women and men work diligently alongside one another, moving fluidly between their roles as workers struggling to keep the power on and soldiers determined to defend the city. One of the last images in the sequence zooms in on the damaged smokestack, standing tall and resilient amid the smoke and rubble, as the mantra of the workers is recited: "*Tất cả cho điện*" (everything for electricity).

The "work-defend-repair" sequence as depicted in the documentary was not unique to the electric plant in Vinh City. By 1972, air strikes on power plants across the DRV had reduced the national energy capacity by more than 70 percent, according to estimates released by the U.S. military.[2] Over a period of ten years, the U.S. air war had doggedly targeted Vietnamese infrastructure

FIGURE 4.1 An aerial image of the destroyed Hàm Rồng (Dragon's Jaw) bridge, 1972. Courtesy of the National Archives and Records Administration.

for annihilation in order to reverse the technical progress of socialist modernization that sustained the war effort in the south. General Curtis LeMay's threat to bomb northern Vietnam "back to the stone age" implied, literally, a strategy of state infrastructural warfare to achieve "forced demodernization" (Graham 2005: 170). Infrastructure networks, called "interdiction points" in U.S. military parlance, especially roads and bridges designed to move large numbers of people, weapons, and supplies to southern battlefields, were hit relentlessly between 1964 and 1968 and again in 1972, while air strikes in the southern panhandle, where cities like Vinh were located, continued uninterruptedly. In the neighboring province of Thanh Hóa, the famous Hàm Rồng (Dragon's Jaw) bridge—a vital transportation link between north and south—was also subjected to hundreds of sorties that would miss or damage the structure, but never completely destroy it.[3] Within days the bridge would be repaired and the passage of vehicles would resume. Over time, the bridge assumed a legendary status as a symbol of Vietnamese resilience and U.S. technological incompetence. In May 1972, in a spectacle of new "smart" warfare technologies, dozens of Phantom bombers pummeled the Dragon's Jaw over a period of two weeks, releasing thousands of pounds of laser-guided bombs that ultimately triggered the collapse of the bridge (figure 4.1).

While both the Hàm Rồng bridge and the smokestack in Vinh became symbolic icons that took on mythical heroic attributes for their perseverance in the

The Current Never Stops   103

face of modern technological warfare, they also emerged as animated and, one might even venture to say, desiring objects (Mitchell 2005) that roused the anxiety of U.S. forces about unruly infrastructures. Of course human action also helped to keep such infrastructure "alive." The motto of the power plant underscored the resolute commitment to the maintenance of technical systems that was expected of worker-soldiers: "Hit us once, we will recover; hit us again, we will again recover; our position is unyielding, we vow to fight until death to keep the current alive and flowing" (Đình 2007: 45). And yet the affective powers imbued in these resilient technological objects intensified even more as infrastructure refused to cooperate and "die." The Air Force had a particularly obsessive relationship with the bridge and became consumed with the task of its destruction. The sublime power that the bridge held over the pilots came to a head in its final demise; its destruction ritually and repeatedly reenacted with excessive force long after its collapse, as if seeking to kill again that which was already dead. In similar acts of iconoclastic violence, planes and ships released thousands of explosives on the smokestack in Vinh in hundreds of attacks, requiring the repair of the plant more than two dozen times (figure 4.2).[4] And yet the smokestack, which seemed to possess a kind of quasi-agency to act and provoke, remained, and it still stands today. As Michael Taussig (1999) has argued, to kill—or seek to kill—an icon, to destroy its affective force and neutralize its power (e.g., its power to move and galvanize passion, desire, and hope) is, paradoxically, to enhance its vitality, making the icon more alive, potent, provocative, and resilient than before.

Taking up the call to examine more closely the affective capacities of infrastructure as political matter at the intersections of technology, materiality, and intimate sociality (Braun and Whatmore 2010), in what follows, I examine the impulses—utopic, electric, and otherwise—embodied in and transmitted by the smokestack, one of the most enduring material forms of urban infrastructure on the landscape in Vinh. Around the world, smokestacks stand as highly visible, and yet abject, relics of industry, whose grand promise of technological prosperity gave way to dystopic realities of risk and calamity. The long-term impact of industry on health and environmental quality continues to make its mark on affected communities, giving shape to an emerging "green energy economy" that trades in renewability as a virtual commodity (Reno 2011). And yet smokestacks continue to hold persuasive power over populations, as observed in contemporary cultural practices of "smokestack tourism" and impassioned efforts to preserve such vertical ruins as landmarks of industrial heritage.[5] Historically, images of smokestacks cir-

FIGURE 4.2 The smokestack and destroyed power plant in Vinh City with the slogan *"Dòng điện không bao giờ tắt"* (The current never stops), n.d. Courtesy of the German National Archives.

culated widely as postcards, personal photographs, press images, and poster graphics, suggesting a range of desires and fantasies about futurity cathected onto these feats of engineering.[6] Clearly there is more to smokestacks than their utilitarian function alone. As objects of affection, smokestacks evoke a range of sensibilities, particularly among those who labored to produce such technological futures: they captivate and repel, embodying hope and despair, especially as they fall into disuse. In postcolonial Vinh, the revolutionary possibility of generating universal electricity for the masses underpinned the collective dream worlds that formed across time and space, also in the face of violent and recurrent disruption and disrepair. The imaginative possibilities of infrastructure that I trace here, particularly those associated with electrification, detracted attention away from what became the deferred dream of socialist modernity. Indeed breakdown itself—from deliberate acts of mass bombing to the violence of state neglect—mobilized the very collective and affective commitments necessary to sustain the promise that energy infrastructure would end the age of darkness and enlighten (literally and metaphorically) the urban population. The resilient brick smokestack, in all its persuasive iterations—signifying seductive ideologies of equality and development, on the one hand, and meaningful social intimacies, on the other—proved essential to this task.

The Spectacle of Infrastructure

In the study of urban infrastructure the promise of progress and prosperity has long been tied to dream worlds of modern technology and an everyday politics of hope in the possibility of securing the "good life." Historically, the notion that public investments in technology and infrastructure growth could result in the transformation and betterment of daily life generated widespread enchantment with technological progress and state developmental projects (Barker 2005; Harvey and Knox 2012). In the United States, widespread fascination with monumental public works, or what David Nye has called the "technological sublime" to identify the profoundly affective, if not transcendental experience of grand technological achievements, can be traced back to the early nineteenth century when the celebration of new democratic technologies that came to dominate the natural landscape became pivotal to national identity (1996: 42–43). Similarly, for Walter Benjamin (1999) spectacular projects of civil engineering in nineteenth-century Europe, made possible by new construction technologies such as iron and glass, radically reshaped sociospatial relations while inspiring dreams of capitalist modernity, mobility, and progress among a desiring urban public. With their capacity to tame unruly landscapes and transform "nature" into "city," spectacular innovations in infrastructure— including a symbolic "urban dowry" of dams, railroads, bridges, and canals— came to signify the technological greatness of a nation, while showcasing the wealth of industrial capitalism (Kaika and Swyngedouw 2000). That this sublime was transmitted to colonial contexts, and even became pivotal to colonial rule, serves as a blunt reminder of how the promise of technological progress, evinced in infrastructure's grand presence, was offered in exchange for political subjectification (Larkin 2008: 245).

During the global cold war, socialist infrastructure, as one of the "cornerstones of state socialist modernity" (Pedersen 2011: 45), likewise sought to convey the emancipatory powers of spectacular public works, and of the Communist Party itself, to an enchanted viewing population. Not unlike the technological sublime of early industrial and colonial capitalism, the monumental projects of high socialism similarly propagated a vision of progressive futurity through the mass production of modern technological systems (see also Boyer, this volume). And yet, arguably, the socialist technological sublime, or "projects of the century" in Soviet terms (Josephson 1995), far exceeded its capitalist counterpart in spectacle and scale. Thousands of new towns and cities with modern industry and infrastructure were rapidly built across the Soviet Union in an effort to transcend class and ethnic difference and unite culturally diverse nations under one centralized governing system. Through the delivery of new technologies to the masses, infrastructure showcased the power and beneficence of the socialist state to provide universal public services, such as heating (Collier 2011). However, as Dimitris Dalakoglou (2012: 572) has shown with Albanian modernization, highway construction, in particular, was more about the spectacle of infrastructure and its allusion to a particular futurity than it was about instrumentality, given the lack of automobility at the time. Similarly, despite frequent breakdowns that I have documented elsewhere (Schwenkel 2015), urban reconstruction in postcolonial Vinh celebrated spectacular feats of civil engineering made possible through the transfer of global expertise and the potential of such infrastructures to create a modern city populated by "new socialist persons" who were adept at using technology in their daily life.

In the postcolonial period, infrastructure as representation became as important to state governance and the management of the population as its function and material form (Larkin 2008: 8; see also Larkin, this volume). This was especially important in postrevolution Vietnam as the state struggled to assert its legitimacy and unite the country around ideas of universal infrastructure embedded in the project of socialist nation building. In posters and billboards, stock images of vertical technologies reaching toward a radiant futurity positioned iconic infrastructural objects—such as industrial cranes, highrise buildings, transmission towers, and brick smokestacks—as backdrops for healthy workers and happy families who stood ready to carry on the revolution (figure 4.3). These scenes of contentment, inclusion, and enlightenment, in stark contrast to prior conditions of French colonial exclusion and underdevelopment, portrayed technology as accessible and beneficial, especially to women and ethnic minorities on the margins of Vietnamese society. In their participation in the project of technological nation building, ethnic minority women, in particular, played a dual role as mothers and protectors of national culture who embraced, and even helped to produce, the fruits of modernity. In one poster, a highland Tai woman in traditional garb, which marks her body as an ethnic Other, stands next to a transmission tower, showing the incorporation of remote areas into the dream of technological progress. In her open hand she holds a green sapling signifying future abundance, while clasping a book on science and technology, stamped with an electron vector, close to her chest. The effect is to merge city and nature, techne and metis, lowlands and highlands, into a unified polity. Similar to postcolonial Indonesia (Barker 2005: 704), technology is shown to transcend class, ethnic, gender, and rural-urban

FIGURE 4.3 "Protect the fruits of the revolution."

divisions in order to affirm the unity of the Vietnamese nation under the leadership of the Communist Party.

This repertoire of images, which continues to circulate widely today, shaped an aesthetics of socialist infrastructure that could be drawn on and deployed in the management of urban populations (Collier 2011: 205). On the one hand, infrastructure visibility and its forms of monumentality were central to techniques of urban governance and agendas for making new socialist citizens, as scholars have noted in other contexts. Caroline Humphrey (2005), for example, has shown how modern construction technologies and morphologies of Soviet architecture were critical to both social and ideological reform. Likewise, Dalakoglou (2012) argues that infrastructure projects were equally about the *social* engineering of new subjects through the labor of construction—and, I would add, the labor of maintenance—as they were about the *civil* engineering of public works. On the other hand, the visibility and performativity of infrastructure, provided it operated as planned, worked to consolidate state power through the public display of the material benefits of socialist citizenship, including the reversal of the uneven development witnessed under colonialism. And yet in Vietnam this emphasis on the spectacle of technological progress is precisely what led to the targeting of infrastructure, especially regional power stations, as the war with the United States escalated.

Electrification and Postcolonial Enlightenment

The early postrevolutionary years in Vietnam marked an optimistic moment in the realization of technological dreams of socialist modernity for the emerging postcolonial state and its population. After nine years of war against France (1945–1954), Vietnam's victory galvanized new and transformative possibilities of state infrastructure and, likewise, new infrastructures of possibility both nationally and internationally. The promise of membership in the block of socialist nations, for example, was contingent on commitment to a particular socialist *political* infrastructure (rule by the Communist Party), and vice versa: the promise of infrastructure was contingent on global political membership, given that aid for industrial development was tethered to formal recognition of Vietnam's statehood. Evolving relations between citizens and the state similarly hinged on such possibilities insofar as the benevolent state (and Party) could demonstrate care and inclusion through the democratic provision of goods and services to those who had labored to build colonial infrastructure— the "coolies" who had revolted and fomented revolution—but had not reaped the benefits of its surpluses. In the immediate years after independence, investment in infrastructure rose steeply as new and refurbished technical systems transformed the built environment in the drive for urban recovery (*khôi phục*). Universal electrification, as a moral and technopolitical project that helped to forge new social attachments and modern sensibilities (Shamir 2011: 13), stood at the crux of this aspiration.

Across Vietnam, the end of the anticolonial war left most urban infrastructure in ruin. Along with French air raids, a scorched-earth policy authorized by the Việt Minh targeted roads, industry, bridges, railways, and buildings that could support French military operations. In the township of Vinh, figures show that the Việt Minh and their supporters destroyed more than 1,300 structures, dug up all access roads (turning them into trenches), and pushed more than 300 railcars and locomotives into the river to thwart the advance of enemy warships (Phạm 2008: 111). In this instance, colonial infrastructure became a tool of tactical resistance, effectively utilizing its materiality against advancing French forces, before the enemy could appropriate and target it for annihilation. In the wake of such destruction, the urban population plummeted from more than thirty thousand inhabitants to a few thousand persons, mostly combatants, as residents evacuated to remote mountainous areas.[7] Light industry was also moved to safer ground. By the end of the war in 1954, the vacant city had no potable water, electricity, or usable roads for transportation. Heavy industry stood in ruins; not one bridge remained intact, according to reports. This landscape of decimated infrastructure would allow for unparalleled opportunities to build new and greater technical systems that could reach a broader scope of the population in both urban and rural locations as part of the promise of socialist modernization.

Infrastructures are often based on speculative calculations (Harvey, this volume). To prevent their obsolescence, they are designed in anticipation of a projected, "not-yet-achieved future" (Gupta, this volume) supported by capacity planning metrics and statistics. The forecasting of infrastructural futures likewise drove rehabilitation efforts in Vinh when the Ministry of Industry approached government officials in Nghệ An about the possibility of constructing the largest power plant in north-central Vietnam. In a letter dated September 3, 1954, the Ministry requisitioned a study to assess anticipated energy demand and output in the provincial capital in relation to existing capacity and predicted patterns of urban growth. This necessitated the frenzied collection of statistical data to evaluate the city's level of "urban civilization" (đô thị văn minh), defined by the ratio between the extent of residual power networks and the percentage of skilled labor among the population, down to the number of street lamps and light bulbs in individual households.[8] Provincial authorities mobilized quickly to respond to the request, but had difficulties locating reliable information in the aftermath of war.[9] In their written response three weeks later on September 23, officials affirmed the modern (hiện đại) orientation of city inhabitants—that they had indeed become familiar with the use of clean and convenient electricity under French rule. While most of the energy generated by colonial public works had been directed toward profit-oriented industry, officials estimated that up to one half of the households in the urban center (numbering three thousand) had electric power in the home, averaging three lights per family.[10] However, because they were obliged to draw on prewar, colonial-era statistics, these figures included wealthier, "foreign"— predominantly French, Chinese, and Indian—administrators and merchants who fled during the resistance. Colonial electrification, in other words, had been a means not to modernize the native Vietnamese, but to consolidate the wealth and power of the ruling elite, not unlike that which occurred in other colonies (see, for example, Chikowero 2007). In all likelihood, the wage workers who totaled more than eight thousand by 1929—mostly peasants who had lost their land to the expansion of colonial industry and other public works, such as airport construction—were not represented in the figure of grid-connected households, given their meager earnings and makeshift dwellings on the margins of the city.[11] As such, one of the seductive claims of postcolonial electrification was to democratize infrastructure and make it accessible to all residents by expanding the network to urban households and rural districts far beyond the former colonial system. Under the banner *"Đảng là ánh sáng"* (The Party is the light), electrification projects brought new legitimacy and meaning to the postcolonial government's moral promise to rescue the population from the darkness of colonization.[12]

For provincial authorities in Nghệ An, undoing colonial power and urban inequality meant rebuilding and reordering the remains of colonial infrastructure, rather than discarding them. In their report to Hanoi, authorities proposed the construction of the electric plant on the ruins of colonial industry and the integration of damaged, but salvageable machinery, such as the Ljungström turbines and generators that technicians were confident they could repair. Hanoi, however, viewed this, and the maximum power capacity of 3,500 kilowatt-hours (kWh), as insufficient for their vision of the scale and speed of modernization. As Leo Coleman (2014: 466) has similarly observed in the case of India, officials in Hanoi were more inclined to invest in new technologies and material infrastructures to demonstrate the political power of the Party and to achieve new levels of industrial productivity, rather than to renovate antiquated systems of a bygone—and no longer modern—era that had enriched France and drained Vietnam of its resources. Moreover, the grid system needed to be large and powerful enough to support the operation of other vital infrastructures in communications and transport. Through its diffusion of electricity, the electric plant was to become the central node and enabler of all other technological productivity in the city, as well as across the wider region.

To carry out its national plan to transform Vinh into one of the largest industrial areas in the DRV, Hanoi turned to international assistance. In the 1950s, the development of political and economic relations with the Soviet Union, China, and other "fraternal" countries grossly expanded the country's infrastructure through the transfer of global technologies that resulted in the refurbishment or new construction of heavy-industry factories, hospitals, housing, and so on. In July 1955, the Soviet Union pledged 400 million rubles in aid to rebuild Vietnamese agriculture and industry with the goal of exporting much-needed foodstuffs and manufactured goods to the Soviet republics (Bích 1983: 18). A percentage of this assistance went toward the new power station in Vinh, which was built under the guidance of Soviet experts and equipped with the latest technologies. Under the Lenin-inspired slogan of *"Điện khí hóa toàn quốc"* (Nationwide electrification), construction on the thermal plant began on June 1, 1956, on the site of the former Société Indochinoise Forestière et des Allumettes (SIFA) industrial complex. SIFA was an early colonial version of today's late capitalist industrial park. Founded in 1907, the grounds housed a match factory and sawmill, one of the most industrious in all of the French colonies, with exports to Hong Kong, Shanghai, and Singapore.[13] A small power plant had been built onsite in 1922 with the capacity to generate 3,500 kWh of electricity for local industry, just as the provincial authorities had noted in their written response to the Ministry. Hanoi's plans were more ambitious, however: workers, cadres, and soldiers, in a grand public spectacle of the people partaking in the global historical project of socialist modernization, worked with Soviet technicians to build a steam generator with a production capacity of 8,000 kWh to power industry, agriculture, government facilities, and public services in Nghệ An and neighboring Hà Tĩnh provinces (Đinh 2007: 45).[14] As such, the power plant came to symbolize a crucial step in the direction of technological advancement with the ability to power the regional economy and, at the same time, to generate trust among the population in the transformative power of global technical innovations.

Against this historical backdrop of infrastructure as a sociotechnical project, unmade and remade under different political regimes, the smokestack in Vinh emerged as an iconic symbol of emancipation from colonial enslavement, backwardness, poverty, and unenlightenment. Across time and space, I argued above, smokestacks have stood as material signifiers of industrial growth and progress; thus the expression "smokestack chasing" from the early twentieth century referred to efforts to modernize small-town America by attracting industry to its doorstep (see note 5). The visual spectacle of soaring brick chimneys with billowing white smoke, and of their physical domination and disruption of landscapes, suggested both economic prosperity and technological prowess: the higher, the wealthier, the more modern and advanced. It is not surprising, then, that a cold war rivalry unfolded around the construction of power plants, with the height of each chimney surpassing the next (the highest in the world was eventually built in Kazakhstan).

In Vinh, the potency and meaning of the smokestack—or *ống khói* (literally, smoke pipe)—were contingent on the state's need for postcolonial infrastructure to be spectacular both in the pace of development and in the application of scientific socialism. It also had to evoke awe if not reverence and to inspire a sense of social cohesion (Nye 1996; Larkin 2008), especially among workers committed to socialist nation building. Still, the materiality and built form of the Soviet-assisted smokestack did not differ so much from its colonial predecessor, nor did it signify a pioneering technology of industry on a new infrastructural landscape. (Its brick construction was somewhat simple and yet sturdy, as its resilience during U.S. air strikes would reveal.) Rather, the smokestack spoke to a new material politics of infrastructure and technique of urban governance *of* and *through* material things (Barry 2013; Lemke 2015). As a biopolitical tool for managing populations (see Anand, this volume; von Schnitzler, this volume), infrastructure makes citizens legible through their encounters with technical systems—for example, in tracking utility usage—allowing the state to penetrate the most intimate spheres of people's everyday lives and monitor their behavior.

Urban political ecologists have shown how the biopolitical work of infrastructure cannot be disassociated from the governance of nature. Through the integrated management of populations and the environment, technical networks transform "the natural" into commodity in the service of urban development (Heynen, Kaika, and Swyngedouw 2006). In contrast to western ideologies that see the biophysical in opposition to the technological (the nature versus city divide), in Vinh the ecological was positioned as symbiotic with the infrastructural, rather than external to it. This reflects a trend in Vietnamese urbanism to view the elemental forces of nature as integral to and constitutive of the city, rather than as rural or green matter out of place. Representations of Vinh's towering smokestack, for example, frequently emplaced the brick structure within a forested landscape at the base of Quyết Mountain, home to the four holy animals (*đất tứ linh*). The smokestack's embeddedness in the sacred land of the nation was affirmed in the iconic image of the power plant released in 1957 that showed Hồ Chí Minh returning to his birthplace of Nghệ An to inspect the construction site alongside Soviet engineers (figure 4.4). The rite of return, combined with the promise of global integration and technological cooperation, marked an important moment in the validation of the DRV after years of snubbing by the Soviet Union. Here, the utopian dream of the developmental state played out in the embrace of global electrification and its potential to uplift the masses to create modern, cultured citizens.[15] As in the Soviet Union (Josephson 1995: 558), technology would serve both the party-state and the people through political and cultural education. Thus, similar to the case of Mongolia that David Sneath has observed (2009: 86), under Vietnamese state socialism, electricity was metonymic for both modern development and individual enlightenment. The civilizing undertones were palpable here and at times mimicked, uncannily, colonial discourse: electricity could shape desirable urban practices through participation in socialist modernity, for example, learning how to properly consume and not waste energy. Similar to water (Anand, this volume; Gandy 2004), electricity would become a key technology for governing the relationship between bodies, moral practice, and the city.

FIGURE 4.4 Hồ Chí Minh meets with Russian advisers during construction of the power plant in Vinh City, June 15, 1957.

The circulation of such images made infrastructure imaginable and desirable for the population. And yet the smokestack was not just a passive icon standing silently in the background. Its presence did something. Among workers, the smokestack provoked emotional responses and attachments because of its association with the end of the violence of colonialism. In a newspaper article based on interviews with workers at the SIFA plant who had revolted in 1930–1931, the Soviet-built smokestack acted as a guarantor of freedom and social justice in contrast to its colonial counterpart. During the revolution, workers had gone on strike to protest SIFA's exploitative practices, such as hiring child laborers and imposing fifteen-hour shifts.[16] In the reportage, the postcolonial smokestack, foregrounded in an image of the power plant to capture its vital presence, was invested with the capacity to free workers from the "slave labor prisons" of the colonial past and to produce a modern, enlightened workforce that assumed ownership of its labor and the factory under socialism.[17] That technological objects like the smokestack served both to liberate workers and to create a new order of socialist persons (*con người xã hội chủ nghĩa*) revealed overlapping and, at times, conflicting state and citizen understandings of the transcendent possibilities of infrastructure: on the one hand was the regulation and cultivation of modern sensibilities through new infrastructural systems, while on the other hand was the fierce desire for independence and escape from colonial (and state) subjugation through broader protections for worker-citizens.

In June 1958, the power plant began operation, with six substations across the two provinces, signifying a new era and geography of electrification. Seven years later, on June 4, 1965, U.S. bombs targeted the factory in two aggressive air raids: one in the morning and a surprise attack in the afternoon that killed eight workers who were attempting to restore power. Another state of emergency ensued, one that would severely constrain the possibility of universal infrastructure as the lights went out once again.

The Intimacies of Infrastructure

The seductive promise of infrastructure lies not with its technical performativity alone, such as the transmission of electricity to users via a more powerfully networked grid. Nor can the promise be understood only in terms of a civilizing project that aspires to create a population of disciplined subjects through cultural enlightenment. There are also important forms of sociality that infrastructures enable that reflect meaningful attachments to the potentiality of technological systems. As Ronan Shamir (2013: 10–11) reminds us, the electric grid is also a *social* assembly that galvanizes collectivities to form (and at times dissolve) around technological objects. Participation in the grand project of infrastructure construction in the postcolonial period, followed by the work of maintenance and repair during the war with the United States, instilled a deep sense of belonging to a greater cause of socialist nation building and national defense in Vinh. Like Brian Larkin has observed of media technicians in Nigeria (2013: 333), in Vietnam, the labor of infrastructure and of working toward a better future made the memory of participation in socialist modernization deeply emotional and transformative. As I will demonstrate, poetry became an intimate way in which people expressed these memories and their lasting solidarities through the iconic figure of the smokestack.

The air war carried out by the United States between 1964 and 1973 reduced the rebuilt city of Vinh to rubble and ash yet again. Unlike the actions of the Việt Minh during the French War, however, here labor was mobilized to restore infrastructure during and immediately after its bombardment, engendering a cyclical routine of "work-defend-repair" that the 1970 documentary film depicted. Through this process, the smokestack took on social and affective meaning beyond its technical function (provide power) and civilizing objective (educate the masses). The postcolonial power plant that had signaled the nation's advance toward global socialism, now under the threat of imperialism, came to stand as a symbol of the resilience of the Vietnamese nation. For workers at the plant, such as Mr. Cảnh and his associates, memories of the war—of the collective labor and traumatic loss of life—became closely entwined with the spirited and resistant smokestack.

I chanced upon Cảnh midway through my fieldwork in 2011 at the poetry board hanging outside a cultural center in social housing in Vinh. As part of a seven-year reconstruction project, East Germany had helped to build the large housing estate with rows of modern apartment blocks and integrated public facilities allocated to civil servants and factory workers. Unlike previous international interventions (singular projects, such as the Soviet-supported power plant, with no fixed master plan), East German material and technical assistance focused on the comprehensive redesign of the city, including the modernization of decimated infrastructure, with the goal of producing a skilled labor force that could enjoy a higher quality of life and work. Cảnh lived in block C4, half of which (forty apartments) had been allocated to workers from the electric plant during the postwar years of centralized distribution of housing. Many of these present-day retirees, Cảnh included, still live there as neighbors. Sporting a French beret, Cảnh proudly showed me his board with poems written in elegant, colorful script according to a six-eight rhyming pattern of interlinking couplets, typical of *lục bát* poetry; a few were accompanied by hand-drawn pictures that illustrated the content of the verse. The poems were carefully arranged and affixed with red tape around a central image of Hồ Chí Minh, under which he had glued a short eulogy from his collection. As I came to know Cảnh and his poetry over the following months, I discovered, like E. Daniel Valentine, that there was a "truth in verse" (2008: 255) that could not be conveyed effectively through prose or more traditional ethnographic methods. Poetry allowed me to see in Cảnh's work (and that of others) an emotional intensity and honesty that, while mediated, was untainted by my own desires to understand the fraught relationship of residents in social housing to technology and the rebuilt environment. It was through poetry that I was able to grasp the range of moral, cultural, and political sensibilities that experiences with urban infrastructure evoked.

Cảnh was one of twelve members of the housing complex's poetry club, which convened weekly to share works in progress and to check that poems followed the rules of iambic verse. Although amateurs, they took their craft seriously. Cảnh, for instance, had been composing poetry as a hobby since he entered the university in 1959, and he was also a member of the Vietnam-UNESCO *Thơ Đường*, or Tang poetry club. The poetic conventions of *thơ đường* were more rigid than that of *lục bát*, but equally sentimental in content. When I first met Cảnh, the club had just finished its meeting at which he had displayed the latest iteration of his board. One poem in particular caught my eye: "*Ống Khói Điện Vinh*" (The smokestack at the electric plant in Vinh), composed on Martyr's Day, July 27, 2008. I was not surprised by the thematic focus; during my research, I had come across a number of songs and poems about infrastructure that celebrated modernity, much as the journalist Ved Mehta recorded at the Soviet-assisted steel plant in India in the 1950s—poems inspired by passionate emotional responses about the development of industry (cited in Coleman 2014: 466–467) and the sacrifice of life it took to achieve it. In Vietnam, this genre of poetry reaffirmed the gendered ordering of infrastructure: Bricks, associated with femininity and women's labor, were compared with the earth (Schwenkel 2013: 266), while in the case of Cảnh's poetry, technology was coded male. The poem "*Ống Khói Điện Vinh*," for example, was handwritten in cursive script and adorned with technological symbols placed in synergy with nature, including an energy transmission tower set amid fields, with a working smokestack at the center of the drawing, positioned next to the factory with birds flying overhead (figure 4.5). In the poem, Cảnh compares the chimney to a flower and its white billowing smoke to the souls of his male coworkers killed on June 4, 1965. "Though the smoke has long dispersed," he wrote of the plant's final closure in 1985 due to chronic breakdown and obsolescence, the smokestack continues to stand as a monument to technological progress, one that also facilitates communion with the souls of the dead who kept their pledge to defend the plant at all costs. "Honorable smokestack," he implored the chimney directly, "help us keep our oaths and traditions alive."[18]

It might be easy to dismiss Cảnh's faith in the developmentalist state and its moral ordering of infrastructure (see also Mains 2012). However, there are a number of double entendres alluded to in his poems that allow for a more equivocal reading. On the surface, the appeal to the smokestack is a clear reference to

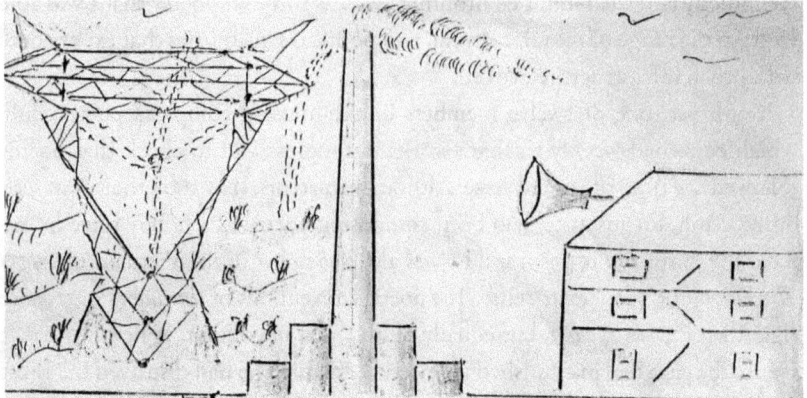

FIGURE 4.5 Poem by Võ Quang Cảnh entitled "*Ống Khói Điện Vinh*" (The smokestack at the electric plant in Vinh).

the vow (*thể*) that all workers took to "fight until death to keep the current flowing"—labor that bestowed upon workers a sense of belonging to the noble cause of national defense. Likewise, "tradition" is an allusion to the state's rhetoric of heroism and the conferral of heroic status on the plant in 1967, as if the smokestack itself had been a soldiering actant in the war.[19] Yet the poem can also be read through the prism of angst about the passing of the socialist dream and changing relationships between society and the state. Through the figure of the revered smokestack, it entreats the paternalistic state to remember its promises of betterment and prosperity. Born in 1941, Cảnh is like many of his generation: He grew up under French rule and survived two wars, one of which he fought in. After studying chemistry in Hanoi from 1959 to 1963, he accepted a job as a low-ranked cadre at the electric plant in Vinh, two years before the first air strike. When the war escalated, he served six years in the military in Area IV, defending the city's industry and port (including the electric station), as well as parts of the Hồ Chí Minh trail. After reunification of the country in 1975, Cảnh returned to the plant, where he lived in collective housing made of wood and thatch (the worker's dormitory had been bombed) until management allocated his family a unit in block c4 of the housing complex in 1981. As was often the case in socialist countries (see, for example, Bray 2005), Cảnh's entire adulthood was spent living and working with the same brigade of people at the power plant (with whom he also fought in the war) at a time in Vietnamese history when state enterprises were responsible for securing basic needs. The influence of the work unit on Cảnh's life, and his attachments to infrastructure, cannot be overstated; it profoundly shaped his identity and poetic sensibilities, as well as his emotional identification with the revolutionary state.

Today, precarity has come to replace stability, however tenuous it was in the postwar years. In the wake of capitalist restructuring, Cảnh's generation has lost many of the social protections it once enjoyed, such as free access to health care, inexpensive utilities, and state-subsidized housing. The socialist contract—infrastructure in exchange for allegiance, Mains writes (2012: 19)—has all but disappeared as prices for public goods and services continue to rise under a new economic model of market socialism. Such conditions notwithstanding, there is a fondness expressed in the poem for the smokestack. It is a respected friend, an object of affection.[20] Even more meaningful, it channels the souls of Cảnh's colleagues, its thick smoke akin to that of incense, connecting the infrastructural world of the living with that of the dead. Like in the newspaper article, in the poem, the smokestack acts to empower collective labor at the plant—including the furnace workers and line installers, as one line reads—and it evokes their shared sense of duty to keep the power flowing. For

Cảnh, the smokestack enabled both military and social action, and it mobilized material solidarities that remain an integral part of his daily life in social housing today.

Through poetry, the smokestack emerged as an object of intimacy that "spurred collective aspirations and became [a site] of ritualization," like grand infrastructural projects elsewhere (Coleman 2014: 468). Beyond the technical and the biopolitical, I was again reminded of the intimate human dimensions of infrastructure one afternoon when Cảnh invited me to his home to share his collection of poems. I arrived at his small, one-room apartment where he lives with his wife to find him tending to his bonsais in the verandah. Cảnh had built a makeshift, wrought-iron extension onto the back balcony, which was common in many units in the five-story buildings and referred to by residents as a "tiger cage." The balcony had become his archive and workroom, and also a space to hang clothes out the window to dry. A reed mat covered the floor, a sitting space where we spent most of my visits. To one side of the room were a number of shelves that held his books and documents. Nearby, next to his hats and ties hanging from a hook on the crumbling wall, was a photograph of General Võ Nguyên Giáp and the national poet, Tố Hữu. He pointed to the image after showing me his books: "I was a soldier; I worked in Giáp's military unit." Giáp had been the commanding strategist who defeated both France and the United States; I asked with curiosity if Cảnh had ever met him. "Of course, many times. I have several poems about him right here," he told me nonchalantly while shuffling through his papers, revealing how his work with, and defense of, energy infrastructures was profoundly moral in terms of his own political and cultural commitments to socialist nation building.

Soon there was a knock at the door and a slight, elderly man in a neatly pressed suit entered smiling. Cảnh jumped up and embraced him excitedly. His name was Tác. They were *đồng nghiệp* (colleagues) who had worked together at the electric plant and had known one other for more than forty years. Tác was in town to visit his younger brother and to organize his mother's death anniversary. Cảnh's home had been his first stop in the city. The three of us sat down on the floor and Cảnh explained what we had been doing. As Cảnh located a series of poems he had written about the electric plant, he described the deep bonds of friendship between him and Tác after surviving years of hardship together at the factory, especially the years of relentless bombing. A poetry reading then commenced. Cảnh began with the poem about the resilient smokestack that I had seen on the poster board at our first meeting. His voice became animated and filled with emotion as he intoned his words for heightened effect, in accordance with the iambic rhythm of *lục bát* (stressed-unstressed-stressed syllable pattern). He read a line—"Whoever comes to the city of Vinh, in the land of Nghệ An"—paused, looked at his friend wistfully, who, deep in thought, nodded and shook his head quietly. At times they touched one another gently as Cảnh recalled painful memories, such as the deaths of their colleagues: "On the fourth of June, the departed have now gone." For these retired electricity workers, the smokestack remained across time a powerfully affective force in their lives, one constitutive of the very materiality of their solidarity (Muehlebach 2017) that bound them together in intimate ways. And yet emotional investments in infrastructure, and the deep sense of comradeship that formed around such attachments, were inspired in this instance not by hope in a brighter, modern future, but by experiences of death and ruination. Here, the power plant, as a site of optimistic infrastructural beginnings and traumatic premature endings, breaks with developmental time and its teleology of progress (Appel, this volume; Gupta, this volume) as technical systems are intentionally struck down and annihilated.

Much has been written about inaugural ceremonies for public works as political rituals that convey to hopeful populations the potential for social transformation (see Harvey, this volume). There is much less attention paid to ceremonies that mark the end of infrastructure—its decommission, dismantling, or destruction. The power plant as a ritualized site of memory making and commemoration was captured in Cảnh's next poem, "*Chuyện kể của chúng ta*" (Our story), which shifted the focus from the smokestack as revered technological object to its role in the annual ceremony that observes the first and deadliest air strike on the plant.[21] Both Tác and Cảnh held hands as Cảnh read unhurriedly:

> The fourth of June has arrived again
> Our friends and colleagues sit together
> Narrating old and recent tales
> Remembering the painful war of past.

Tác's eyes welled with tears as he looked off into the distance, absorbed by the cadence of the uttered words. Cảnh lyricized his personal injuries from the bombing, which he described as preferable to the fate of comrades who never returned. Arms around one another, Cảnh brought the poem to a close:

> Heavy emotions, deep meanings
> The aching trace of bombs
> As we depart, we remember the day, the spirits,
> And our oath to die for the current.

The pair sat silently for a moment until Cảnh moved to locate a photograph, which he handed to me. In it, Cảnh stood in front of an audience reading this very poem at the June 4 ceremony that he attends every year with Tác and his neighbors from the plant. "Many people cried when I read it," he said, showing how the sociopolitical effects of violence carried out against infrastructure continue to resonate emotionally among his generation.

Performative enactments around infrastructure, including the poetry readings and the commemorative ceremonies, remain essential to the identity of Cảnh and his colleagues, and they continue to shape their everyday activities. After a few additional readings (about mothers, growing old, and youth brigades), Cảnh grabbed his yellow hard hat he still keeps from the plant, which hangs next to his ties, and headed outside to park motorbikes. It was late afternoon and the outdoor market would soon be buzzing with shoppers on their way home from work. A small group of retired coworkers, also in their hard hats, was roping off an area where people could park their bikes for 3,000 Vietnamese đồng (US$0.15), money that the team would use to supplement their monthly pensions of US$80. Tác said goodbye and Cảnh got to work with the rest of the brigade, while I sat on the sidelines listening to the men (and one woman) complain about inflation and the compulsory privatization of their apartments. Long after the deactivation of the smokestack, the intimacies that had formed through their participation in infrastructure—in the production, defense, and maintenance of electric power—remained remarkably intact, even strengthened, under present conditions of economic vulnerability.

Conclusion: On the Obsolescence of Infrastructure

"Railroad tracks," with the peculiar and unmistakable dream world that attaches to them, are a very impressive example of just how great the natural symbolic power of technological innovation can be. —WALTER BENJAMIN (1999)

———————

Like the dreams of technological mobility attached to Benjamin's wrought-iron tracks of modern railways, hope and aspiration for a progressive modernity attached themselves to the postcolonial power plant and the icon of the towering smokestack in Vinh, symbolizing the revolutionary potential for universal electricity and a release from the violence of infrastructures past. The different forms of social and political potency that the smokestack assumed across cycles of ruination and reconstruction compelled emotional attachments to infrastructure that proved useful to the state and its foundational myths, as seen in the case of Cảnh's unwavering (though not uncritical) loyalty. And yet his performative enactments also showed that infrastructure is more than merely the stuff of political matter. As Antina von Schnitzler has observed with water meters, technological objects are "open to a diverse set of ethical claims and affective investments" (2013: 682). And yet the possibilities of infrastructure in Vinh, as in South Africa (see also von Schnitzler, this volume), have been unpredictable and contingent: on the one hand was the emancipatory potential to bring an end to poverty, inequality, and "backwardness" through inclusion in socialist modernity, and on the other hand a risk of violence and death emerged in response to these very efforts.

If infrastructure is an ongoing process of becoming as Stephen Graham (2010) and others have noted, then more attention needs to be paid to its technical and sociomaterial histories, that is, to its distinct object biography (and afterlife, Benjamin would argue) across time and space. Members of Cảnh's generation, for example, have lived through multiples iterations of urban infrastructure under colonial capitalism, cold war imperialism, socialist internationalism, and neoliberal marketization that reveal the back-and-forth movement of urban technical systems—their stasis, undoing, and remaking—as constitutive of what Appel (this volume) identifies as "infrastructural time."

Infrastructure has a tendency to become obsolescent quickly. In 1985, the electric plant in Vinh City definitively shut down for this very reason. After its bombardment, the plant produced insufficient energy for a growing population, too little for the state to fulfill its promise of technological advancement. Inevitably, the current *did* stop. Damaged turbines kept the city dark until 1977. Imported street lamps were merely ornamental, offering a spectral façade of modernization, much like Albania's highways. In social housing, new technical systems remained inoperable. The expansive infrastructure built by East Germany was impossible to sustain without a fully functional and operational power grid.

The subsequent disassembling of unfeasible infrastructure began soon after East German engineers left in 1981. During my fieldwork thirty years later, the local government embarked on a project to reinstall street and building lights within the housing complex. Each household, including mine, was required to contribute 100,000 đồng (US$5). Văn, a neighbor, was one of many residents to complain about these obligatory "donations." One gray December morning as we sat outside drinking tea, he reflected on the ephemeral moment of postwar enlightenment:

You know, when [East] Germany built these homes, we had electricity and lights on the buildings. The streets were lit up! We didn't have to pay

FIGURE 4.6 The smokestack as a national heritage site, 2015. Photo by the author.

anything. And then the state came and *cắt đi* [cut it all]—they turned off the lights and removed them! Even today, it's still dark. Why? *Tiết kiệm*—they wanted to save money! Everybody groaned when the state undid the electricity. Today we have to contribute to everything: fixing the roads, paying for [building] repairs, donating for floods and Agent Orange . . . and now we have to pay collectively to reinstall the lights. This is senseless! My pension is 1.4 million đồng [US$70]. How should I afford all of this?[22]

Văn's words captured the growing bitterness I observed among residents as the burden of infrastructure and its maintenance shifted to their limited pocketbooks, a change that was felt to be a betrayal of the socialist contract: state care for loyalty and sacrifice. Today the rising cost of energy is a source of much anxiety for poorer residents, which motivates people like Cảnh and his brigade to work for extra money in their old age. Electric bills that typically ranged between US$2 and US$5 increased more than 10 percent over the course of my research during a period of rapid inflation tied to the global financial crisis. Meanwhile, older residents like Văn and Cảnh, who had entered into a reciprocal relationship with the state through their labor and military service, wondered why they had to pay for utilities at all: How could their basic needs, in other words, become marketized?

With the retirement of the power plant, the smokestack came to assume a new role in the city. In 1997, the Ministry of Culture and Education bestowed the status of national heritage on the site, twelve years after its closure (figure 4.6). This recognition imbued the historic chimney with new political and technological significance as industrial heritage owing to its contributions to both the development *and* defense of the nation (see also Muehlebach 2017). And yet, as I have argued here, there was much more to the material life of this resilient infrastructural system. Beyond the slogans, wires, dispatchers, turbines, mountains, foreign experts, and bombs that mobilized labor solidarity lay the social intimacies and sensibilities that the smokestack had mediated. In its new role as a protected national monument deemed worthy of remembrance, the smokestack reveals its tales of history through annual ritual ceremonies (at the site's altar), through tourism (along with a new exhibition house), and through literary devices, such as Cảnh's poetry that transmutes the sacred object into instruments of nature: a flower in one poem, a bird in another. Obsolescent infrastructures, as sanctified ruins (and objects of touristic desire), can remain affectively charged long after the technological dreams attached to them have vanished, in this case linking life and death, and nature and city, while sustaining the social collectivity through memories of the infrastructural legacies of past futures.

NOTES

1 This scene marks an important moment in the gendering of infrastructure, which is usually associated with the domain of male power and expertise. On the role of women in the construction and breakdown of infrastructure, see Schwenkel (2013).

2 July 3, 1972, Memorandum for the President, page 5, U.S. National Archives and Records Administration (NARA), Nixon Presidential Library. National Security Files, Vietnam Subject Box 96. Folder: Air Activity in Southeast Asia, January–August 1972, Volume 3.

3 A 1967 CIA report on the effectiveness of bombing found that only one in 47 bombs hit a targeted bridge. "Air Effort against North Vietnam." Memorandum for Henry

Kissinger, September 5, 1972. NARA Nixon Presidential Library, Vietnam Subject File, Box 97, Folder 4: Air Activity in Southeast Asia Vol. IV, Sept-Dec 1972 [Folder 1 of 2].

4 Standard figures report that the plant was subjected to three hundred air strikes that released an estimated 2,319 bombs and 149 missiles, requiring its repair twenty-six times (Bùi 1984: 158).

5 In travel discourse, smokestack tourism is also referred to as "chasing smokestacks," which is not to be confused with the economic development strategy in the 1930s called "smokestack chasing," which saw the potential for growth in attracting industry to small towns. Smokestacks thus came to be associated with prosperity and material betterment.

6 Such images continue to hold persuasive power today in diverse ways, from the Anaconda Smoke Stack State Park in Montana to the German board game *Power Grid*.

7 Six months after the end of the war in 1954, the population was recorded by the Municipal People's Committee at 7,650 persons.

8 File 575, Tài liệu về xây dựng nhà máy điện Vinh, năm 1954 [Documents on the construction of the power plant in Vinh, 1954], Nghệ An Provincial Archives.

9 Drafts of the letter show handwritten notes stating "*chưa rõ*" (not yet clear) following much of the requested information. See File 575.

10 Handwritten drafts, on the other hand, estimated only 1,500 households, or 25 percent of the population. It is possible that officials chose the higher number to convey to Hanoi the technological possibilities of infrastructure growth in Vinh, which is located in a region historically known for being poor and underdeveloped.

11 Their earnings were further reduced by taxes and fees for workplace amenities, making electrification highly unlikely.

12 Thank you to Phạm Phương Chi for this reference.

13 A Swiss entrepreneur had founded the matchstick factory as a small operation around 1891 (Bùi 1984: 152). Although the French colonized Nghệ An in 1885, they did not invest in any large-scale industrial and infrastructure development until after 1900 (Nguyễn 2008: 134).

14 See also Bùi (1984: 152) on the amount of human labor required to collect construction materials (such as stone), level the site, and build the plant—labor that was essential to producing socialist subjectivities as Dalakoglou (2012) has argued, through a sense of collective participation in the project of modernization.

15 Vu Can (1975: 155–57) recounts a similar story of uplift among ethnic minorities who opposed the construction of a hydroelectric dam in 1962 on account of the evil spirits that haunted the river that was to be harnessed. They eventually acquiesced and on inauguration day, when the magnificent light illuminated their villages at night, science had indeed won over superstition.

16 There were more than 750 workers at the factories, the majority of whom were women and children.

17 "Ba mươi tư năm ấy đổi thay đã nhiều" [Much has changed over the past thirty-four years], *Báo Nghệ An* [Nghệ An Press], September 12, 1964, pp. 2, 4.

18 All translations are the author's.

19 For example, the emphasis in state discourse to maintain the nation's "heroic tradition," or *truyền thống anh hùng*.

20 Reverence for the smokestack was also a theme in state discourse. Another chimney-centric slogan during the war was "The chimney is down, we built that chimney with respect; if the respected chimney is down, we will build smoke passages underground."

21 Thirteen workers were "sacrificed for the current" (*hy sinh vì dòng điện*): eight on June 4, 1965; one—two weeks later—on June 17; one on February 4, 1966; two on April 16, 1968, and another on April 18, 1968. A monument and altar to those killed were built onsite in 2007 next to the smokestack, marking a new chapter in the social life of infrastructure as national and industrial heritage.

22 Interview, Vinh City, December 2, 2010.

REFERENCES

Barker, Joshua. 2005. Engineers and political dreams: Indonesia in the satellite age. *Current Anthropology* 46(5): 703–727.

Barry, Andrew. 2013. *Material politics: Disputes along the pipeline.* New York: Wiley-Blackwell.

Benjamin, Walter. 1999. *The Arcades Project,* trans. Howard Eiland and Kevin McLaughlin. Cambridge, MA: Harvard University Press.

Bích Sơn Hoàng. 1983. *Việt Nam, Liên Xô: 30 năm quan hệ (1950–1980)* [Vietnam, Soviet Union: 30 years of relations]. Hanoi: Ngoại Giao.

Braun, Bruce, and Sarah J. Whatmore, eds. 2010. The stuff of politics: An introduction. In *Political matter: Technoscience, democracy and public life,* ed. Bruce Braun and Sarah J. Whatmore, ix–xl. Minneapolis: University of Minnesota Press.

Bray, David. 2005. *Social space and governance in urban China: The Danwei system from origins to urban reform.* Stanford, CA: Stanford University Press.

Bùi Thiết. 1984. *Vinh—Bến Thuỷ.* Hanoi: Văn Hóa.

Chikowero, Moses. 2007. Subalternating currents: Electrification and power politics in Bulawayo, colonial Zimbabwe, 1894–1939. *Journal of Southern African Studies* 33(2): 287–306.

Coleman, Leo. 2014. Infrastructure and interpretation: Meters, dams, and state imagination in Scotland and India. *American Ethnologist* 41(3): 457–472.

Collier, Stephen J. 2011. *Post-Soviet social: Neoliberalism, social modernity, biopolitics.* Princeton, NJ: Princeton University Press.

Dalakoglou Dimitris. 2012. "The road from capitalism to capitalism": Infrastructures of (post)socialism in Albania. *Mobilities* 7(4): 571–586.

Daniel, E. Valentine. 2008. The coolie. *Cultural Anthropology* 23(2): 254–278.

Đình Phượng. 2007. Nhà máy điện Vinh trở thành di tích lịch sử quốc gia [Vinh electric plant recognized as a national heritage site]. *Điện Lực* [Electricity] 12: 45, 59.

Gandy, Matthew. 2004. Rethinking urban metabolism: Water, space and the modern city. *City* 8(3): 363–379.

Graham, Stephen. 2005. Switching cities off: Urban infrastructure and US air power. *City* 9(2): 169–193.

Graham, Stephen, ed. 2010. *Disrupted cities: When infrastructure fails.* New York: Routledge.

Harvey, Penny, and Hannah Knox. 2012. The enchantments of infrastructure. *Mobilities* 7(4): 521–536.

Heynen, Nik, Maria Kaika, and Erik Swyngedouw, eds. 2006. *In the nature of cities: Urban political ecology and the politics of urban metabolism.* New York: Routledge.

Humphrey, Caroline. 2005. Ideology in infrastructure: Architecture in Soviet imagination. *Journal of the Royal Anthropological Institute* 11(1): 39–58.

Josephson, Paul R. 1995. "Projects of the century" in Soviet history: Large-scale technologies from Lenin to Gorbachev. *Technology and Culture* 36(3): 519–559.

Kaika, Maria, and Erik Swyngedouw. 2000. Fetishizing the modern city: The phantasmagoria of urban technological networks. *International Journal of Urban and Regional Research* 24(1): 120–138.

Larkin, Brian. 2008. *Signal and noise: Media, infrastructure, and urban culture in Nigeria.* Durham: Duke University Press.

———. 2013. The politics and poetics of infrastructure. *Annual Review of Anthropology* 42: 327–343.

Lemke, Thomas. 2015. New materialisms: Foucault and "the government of things." *Theory, Culture and Society* 32(4): 3–25.

Mains, Daniel. 2012. Blackouts and progress: Privatization, infrastructure, and a developmentalist state in Jimma, Ethiopia. *Cultural Anthropology* 27(1): 3–27.

Mitchell, W. J. T. 2005. *What do pictures want? The lives and loves of images.* Chicago: University of Chicago Press.

Muehlebach, Andrea. 2017. The body of solidarity: Heritage, memory, and materiality in post-industrial Italy. *Comparative Studies in Society and History* 59(1): 96–126.

Nguyễn Quang Hồng. 2008. *Kinh tế Nghệ An từ năm 1885 đến năm 1945* [The economy of Nghệ An from 1885 to 1945]. Vinh: Lý Luận Quốc Gia.

Nye, David E. 1996. *American technological sublime.* Cambridge, MA: MIT Press.

Pedersen, Morten Axel. 2011. *Not quite shamans: Spirit worlds and political lives in northern Mongolia.* Ithaca, NY: Cornell University Press.

Phạm Xuân Cẩn. 2008. *Văn hóa Đô thị với Thực tiến Thành phố Vinh* [The practice of urban culture in Vinh City]. Vinh: Nghệ An.

Reno, Joshua. 2011. Motivated markets: Instruments and ideologies of clean energy in the United Kingdom. *Cultural Anthropology* 26(3): 389–413.

Schwenkel, Christina. 2013. Post/socialist affect: Ruination and reconstruction of the nation in urban Vietnam. *Cultural Anthropology* 28(2): 252–277.

———. 2015. Spectacular infrastructure and its breakdown in socialist Vietnam. *American Ethnologist* 42(3): 520–534.

Scott, James C. 2009. *The art of not being governed: An anarchist history of upland Southeast Asia.* New Haven, CT: Yale University Press.

Shamir, Ronan. 2013. *Current flow: The electrification of Palestine.* Stanford, CA: Stanford University Press.

Sneath, David. 2009. Reading the signs by Lenin's light: Development, divination and metonymic fields in Mongolia. *Ethnos* 74(1): 72–90.

Taussig, Michael. 1999. *Defacement: Public secrecy and the labor of the negative.* Stanford, CA: Stanford University Press.

von Schnitzler, Antina. 2013. Traveling technologies: Infrastructure, ethical regimes, and the materiality of politics in South Africa. *Cultural Anthropology* 28(4): 670–693.

Vu Can. 1975. *North Viet Nam: A daily resistance.* Hanoi: Foreign Languages Publishing House.

PART II. *Politics*

5

Infrastructure, Apartheid Technopolitics, and Temporalities of "Transition"

ANTINA VON SCHNITZLER

In a speech given in July 2013 by Julius Malema, the controversial former African National Congress (ANC) Youth League president and founder of a new opposition party, the Economic Freedom Fighters, toilets took center stage. In the widely publicized speech, Malema decried the fact that, twenty years after the end of apartheid, many residents of townships and informal settlements still lacked access to proper sanitation. "You fail to deliver toilets and you think we can take you seriously," he said, addressing the government. "We need toilets, you must never be ashamed. If there is a need for a toilet revolution we must engage in a toilet revolution. Let them say we are a toilet organisation."[1] Malema's statement was only the latest in a series of highly mediatized debates and demonstrations in which toilets had taken center stage. In the months leading up to his speech, residents of informal settlements around Cape Town had repeatedly emptied buckets of sewage at central locations in what came to be known as the "poo wars" in protest of the temporary toilets that the municipality had set up in many informal settlements.

By that point though, toilets had already been firmly lodged within the Capetonian public imagination. Two years earlier, the local elections had been infamously preoccupied with a different set of toilets. During what came to be known as the "the toilet election," the ANC had accused the locally ruling Democratic Alliance (DA) of building toilets without enclosures in the townships.[2] These so-called open toilets were part of a larger "site and service" scheme in which residents of informal settlements are provided with plots of land and infrastructure connections, but are expected to build their own housing. Thus, the toilet walls too were to be added by residents themselves. In articles framed by photographs of the toilets sitting exposed on empty plots, political commentators lamented the "sight of a woman sitting on a toilet" as an indignity of incomparable proportions. Soon more sarcastic talk of "open air toilets" or "cabriolet toilets" filled the press. As the *Mail and Guardian* put it dryly, "The municipal election campaign is in the toilet, literally."[3]

The public spectacles and the moral anxieties prompted by the toilets raised much larger questions about democracy after apartheid; here, as in many other instances, infrastructure became a medium for the expression of a diversity of disappointments with the postapartheid condition. Many established political commentators later suggested that the open toilet debacle had been an "election stunt," especially since it turned out that the installation of open toilets was similarly practiced in ANC-ruled municipalities. There was broad agreement that while the toilet situation was indeed dire, the way in which the issue had monopolized the election campaign and banalized public debate in the country was a sign of a broader, and largely worrisome, populist turn. Throughout the commentary, an image thus emerged of democracy under threat from ruthless populists whose strategy of spectacle and emotive sloganeering degraded a cleaner, more sober, and deliberative ideal of democracy.[4] Indeed, Malema's performative call for a "toilet revolution" was provocative precisely against this normative backdrop of a purer, more immaterial political that *should be* beyond the private realm of bodily necessity. Here, again, the toilets were of crucial importance, separating democrats from demagogues, civil from uncivil society.

The rise of the toilet in South African politics and the public imagination obviously provides multiple themes that could interest anthropologists, from the moral economies of dirt and dignity it evoked to the Rabelaisian politics of the "poo wars" that continued to make headlines for months.[5] In this chapter, my aim will be more limited. I want to take the toilet election as a point of departure for some reflections on democracy and the politics of infrastructure in South Africa, in the hope of thinking more broadly about the possibilities and limits of an anthropology of infrastructure. More specifically, and inspired by the theme of this collection, I am interested in exploring what epistemological promise infrastructure may hold for anthropology.

As Brian Larkin (2013: 375) has recently suggested, infrastructures work at numerous levels at the same time. They are functional, in that they enable, constrain, and shape the circulation of goods, people, or ideas, but they can also become signs and symbols. They can make us do things, and elicit affect, and they can work via the spectacular and the sublime as much as the mundane and the banal. They can also become ethical objects, conduits of power, or be wielded as political tools. It is this multilayeredness and multivalence that, of course, makes them such productive objects of ethnographic inquiry.

But in order to ascertain the stakes of our inquiries, the interventions they mobilize, and the critical debates they intervene into, we might also need to articulate in more precise terms what is aspired to, what is gained, and what is foreclosed in a particular choice of ethnographic object. With apologies to Ian Hacking (1999), we might ask "what's the point" of an anthropology of infrastructure? With David Scott (2005), who provocatively takes up Hacking's question, we could further ask, what is its upshot and conceptual or critical space of intervention?

I would like to propose that the perspective of infrastructure—including an attention to form—may offer a useful vantage point onto democracy and a more expansive theory and vision of what it means to act politically in the postcolony and beyond. More specifically, it may provide an account of the political that is attuned to the material, affective, counterpublic, or indeed nonpublic *forms* of political engagement. In turn, it may thus enable us to contribute to existing efforts to rethink normative liberal-secular accounts of politics with their imaginaries of the public as a disembodied sphere of deliberation, fantasies of "free" circulation, and global modularity.[6] In this chapter, I want to conceptually sketch some of the forms of the political that infrastructures enable with a particular focus on sociotechnical, material modalities of politics.

While the toilet spectacle pointed to the ways in which infrastructures became a "matter of concern" that provoked affective responses (from sympathy to compassion and outrage), in the following, I track a less visible politics of infrastructure, what one might term a politics of nonpublics.[7] Here, the boundaries separating the political and the administrative are more murky, akin to the account Partha Chatterjee (2004, 2011) provides of "political society" in which administrative connections to the state become the sites where rules and norms can be stretched and bent, and where space opens for a politics that has been foreclosed in the formal sphere of politics.

If one of the contributions of studying infrastructure is to complicate accounts of "free" circulation—showing, for example, that the space for the political is always already shaped and performed by forces or mechanisms that often remain invisible—here I want to focus on the ways in which the political may also take shape *at the registers and in the forms of the infrastructural*.[8] This is a less visible politics that does not necessarily circulate freely, that may appear constrained, and in which liberal norms do not necessarily govern. I am concerned here, in other words, with a technopolitics in which infrastructure itself becomes a modality of political action.[9]

Just after South Africa's first democratic elections, Ivor Chipkin suggested in a short paper that South Africa is defined by a peculiar, differentiated citizenship based on what he called the "non-identity between the national and the local state" (1995: 38). There was, he suggested, a sharp disjuncture between experiences of national and urban citizenship, a disjuncture shaped both by apartheid rule and by the specific forms in which the antiapartheid struggle unfolded. This disjuncture and special status of urban citizenship in South Africa, Chipkin presciently suggested, would be important in the years to come.[10] Nearly twenty years later, Zwelethu Jolobe (2014) raises a strikingly similar concern. Jolobe notes that despite the many changes that have taken place at the local government level, including not least deracialization, one aspect that has not changed much is the underlying continuity of "top-down structures" that, he argues, are in important respects continuous with the colonial and apartheid eras. Note that these analyses are not primarily focused on concerns that are often raised in public debates in South Africa in this context, such as "lack of capacity," corruption, or problems of what is routinely referred to as "delivery." Rather, both Chipkin and Jolobe speak to apartheid legacies with a more durable if often less visible foundation. They also point us to the multiplicity of *experiences* of the state and, indeed, of South Africa's globally celebrated "transition."[11]

This focus on the nonidentity of the national and the local also raises the question of how we account for and render legible apartheid's less visible legacies and their material and affective durability, what one might, with Ann Stoler, term apartheid's debris.[12] How do we capture what Stoler terms "the evasive space of imperial formations . . . as well as the perceptions and practices by which people are forced to reckon with features of those formations in which they remain vividly and imperceptibly bound" (2013: 8)? Indeed, if the reference point of South Africa's "transition" was always already national and its periodization rendered in national, homogeneous time, how do we write the multiplicity of local transformations, their "uneven pace" and heterogeneous temporalities?[13]

Here, I would like to suggest that the infrastructural provides a register of inquiry and a vantage point onto these disjunctures of democracy, as well as a different way of thinking about the periodization and multiple temporalities of "transition."

THE CAPE TOWN toilet debate followed on the heels of a decade during which South Africa's "miracle transition" narrative had been interrupted by waves of, at times, violent protests concerned with a diversity of grievances, from housing or the provision of basic services such as water or electricity to boundary disputes or internal party political battles.[14] These so-called service delivery protests have often been received by the media and the public with much less sympathy. The sheer number of such protests has recently led commentators, no doubt slightly hyperbolically, to call South Africa the "protest capital" of the world. Less visibly, the postapartheid period has been defined by widespread illicit acts, such as the nonpayment of service charges; the bypassing, tampering, or destruction of water and electricity meters; and illegal connections to services.[15] While the contemporary protests and popular illegalities uncannily resemble the tactics of the antiapartheid struggle, today, they lack the political languages of liberation that authorized and gave them meaning in the past. In the absence of such larger narratives, nonpayment, illegal reconnections, and violent protests are often framed by officials and the media as irrational, amoral, or criminal acts. Thus, they become the responsibility of the courts or a diverse array of experts, including engineers, utility officials, or local bureaucrats. In contrast with the late apartheid period during which the liberation movement often authorized and subsumed such acts within a nationalist language of struggle, today such acts appear increasingly spectral: senseless acts produced by faceless mobs and often defined by violent irrationality or Hobbesian self-interest. At the same time, concerns that were central to the antiapartheid struggle are now often "transduced" to an infrastructural terrain where they continue albeit in a different language, one that in the present is for the most part no longer intelligible as "speech."[16]

In what remains of this chapter, I would like to provide some reflections on the forms that this politics of infrastructure takes by first sketching a brief genealogy of apartheid technopolitics and, second, by giving an example of the *kinds* of infrastructural politics at work in South Africa today.

Apartheid Technopolitics

In several ways, apartheid as a political project depended upon and was conjured into being by specific infrastructural modalities of power. This was so particularly in the urban areas, in which, following the ideology of grand apartheid, black residents were conceived as "temporary sojourners" whose permanent home and political representation were envisioned to ultimately be in the rural Bantustans. Here, infrastructures became symbols, conduits and forms of power, but they also shaped habits and the senses. This combination of the symbolic, the biopolitical, and the affective produced a very specific political terrain, one whose remainders shape the contemporary politics of infrastructure. Crucially, and unlike in places where the extension of infrastructure was often closely bound up with citizenship, for the most part apartheid infrastructures worked to prohibit the emergence of a (nonracial) public.[17]

First, and most visibly, apartheid was infamously symbolized via its infrastructures. Think of the segregated public transport and amenities, the jarring images of race-specific entrances or "whites only" benches that came to metonymically represent the injustices of apartheid. Infrastructures and access to them also became subject to juridical regulation by the battery of racial legislation passed in the aftermath of 1948. The most symbolically charged of these was the Reservation of Separate Amenities Act of 1953. Beyond the symbolic, apartheid was also made functional via its infrastructures insofar as it was at one level simply a grand scheme to channel and police mobility. Indeed, it could be argued that apartheid (qua "separate development") was precisely *about* infrastructures—institutional separation meant the use of separate infrastructures (from schools, to transport, public spaces, etc.) and the production of a racial economy. The latter could not have come into being without the infrastructure of the pass law system by which millions of people were channeled via labor bureaus, passbooks, and rail tracks to factories, mines, and farms to scientifically manage labor supply. This system, in turn, was predicated on innumerable technologies, large and small.

Apartheid, as a racialized state project built on what Stephen Gelb and John Saul (1981) called "racial Fordism," aspired to create a thoroughly engineered society, founded on a trust in numbers and technoscientific expertise.[18] Indeed, though such techniques did not always produce the desired effects, apartheid was unthinkable without the administrative power of large parastatals (Clark 1994), the technical forms of accounting and measurement that defined the labor bureau system (Hindson 1987; Posel 1991), biometric identification (Breckenridge 2014), and the technoscience of race and identity on the basis of which groups were often created and managed (Dubow 1995; Bowker and Star 2000; Edwards and Hecht 2010). In a great variety of ways, apartheid infrastructures thus worked not merely to enable circulation (which is what we often think infrastructures are primarily designed to do), but also to impede, prescribe, and prompt movement.

As Ivan Evans (1997) argued in an important book, apartheid's form of rule was distinguished from its segregationist predecessor by its focus on the administrative powers of the local state. In the urban areas more specifically, Evans pointed to the centrality of infrastructure in containing black opposition. After the appointment of Hendrik Verwoerd as minister of native affairs, "native administrators became deeply enmeshed in providing cheap, mass-produced housing, public utilities, and mass transport to the African working class," which in turn was central to "disorganizing African opposition" (Evans 1997: 7).[19] This was especially true in urban areas, where native policy was exercised primarily through civil administration: "collecting rents, supplying public utilities, and planning transport routes" (9).

With the myriad ways in which this apartheid technopolitics propped up this system also came particular ways of being in the world.[20] In the absence of political rights, township residents experienced the state primarily via administrative connections. The state was made manifest in the most spectacular fashion through its repressive policing powers and the necropolitics it produced (Mbembe 2003). And yet, in its various guises, the state was also a landlord for plots and houses, a provider of infrastructures and basic services, and a collector of payments and rents.[21] Rather than merely racist ideology or a system of repression, then, apartheid also produced much more mundane modes of being in the world, including a set of habits, sensory experiences, and forms of sociality. As Jacob Dlamini (2009) suggests, being in the urban areas was also a "felt experience" that in turn produced "embodied memories," in particular for the early African migrants to the city who "spoke of [the urban experience] as a series of new sounds, smells, textures, tastes and sights" (129). Similarly, infrastructures and associated bureaucratic procedures played a central part in shaping these more sensory and embodied experiences of state power.

If apartheid infrastructures had as one of their primary goals to prevent the emergence of a public of sorts, this was most obvious in the mass building of townships from the 1950s onward as spaces intentionally without important city features. Conceived as mere dormitories for laborers and built far away from the white city centers, the townships had no plazas or public squares and no businesses were allowed to operate. Phil Bonner and Lauren Segal's depiction of the newly built areas during the 1950s vividly portrays how infrastructures worked on the senses: "Row upon row of identical dirt streets radiating from a central hub, line upon line of drab, cheap, uniform houses, a colour-less, mind-numbing monotony. It is almost as if the government felt that through regimentation and uniformity it could establish a firmer control that could not be challenged" (1998: 34). This experience of uniformity was reinforced by the absence of street names and addresses. Until the 1990s, a dwelling was indicated only by a number and the township, clearly marking the townships as camplike "spaces" rather than urban places made up of singular forms and distinctive parts.[22]

Instead of building and supporting a public, infrastructures often followed a security or military logic. The grids of streets were planned such that they could be easily surveilled and closed off. In many townships, radial roads led to spaces built specifically to be used as potential weapon arsenals in case of protests. Similarly, electricity cables were first extended to the townships to service tall floodlight poles and enable surveillance at night. Only later did township residents receive electricity within their homes. Thus here infrastructures were used not to produce a public (let alone a postapartheid nation!), but, on the contrary, to prevent it from coming into being. These were decidedly illiberal forms of infrastructure.

And, to some extent, if in a very different way, this was also true for the wealthier white suburbs, which were also, if of course much less severely, marked by the absence of certain public infrastructures. While the apartheid state invested heavily in white areas, here, infrastructures were often de facto privatized. Given the reliance on cars, there was often less focus on public transport, streetlights, or pavements. Parks and other public spaces were relatively unimportant, because most whites had access to and preferred the safety of backyards, or indeed large gardens or private country clubs. And yet even in the absence of such infrastructures, the streets became locations for domestic workers to meet and socialize during break times, and this is how they are used in many wealthier, historically white suburbs to this day. Indeed, many of these separations introduced in the apartheid era continue, in part because of the intransigence of infrastructure.[23]

Given these multiple ways in which infrastructure symbolized, produced, and secured apartheid, and given the absence of a legitimate formal political sphere, it is unsurprising that the antiapartheid struggle often unfolded on similarly infrastructural terrain. While the liberation movement became known to the outside world for its nationalist demands for political rights, it was often made up of localized struggles, many of which took infrastructure as an object and terrain of struggle in a wide variety of forms. Through actions by Umkhonto we Sizwe, the ANC's armed wing, such as the bombings of rail tracks, power plants, or, in a particularly spectacular incident, the Sasol oil refinery, infrastructure became a central part of a symbolic form of guerrilla warfare. Somewhat less spectacularly, but in many ways more effectively, the mass burning of passbooks, one of the most iconic instantiations of *grand apartheid,* similarly operated on this symbolic plane.

The prominence of infrastructure in the struggle increased dramatically during the 1980s, not least because infrastructure became ever more closely linked to the security and armed forces. In the context of the rent boycotts during which residents in townships all over South Africa stopped paying rent and service charges, material and fiscal connections to the state became a primary terrain of struggle. Evictions for nonpayment of services became the sites of militant mobilization, with activists often letting residents back into their houses as a form of protest. Activists circulated petitions to contest the accuracy of electricity billing. Residents carried night-soil buckets and emptied them in front of council offices. Mass meetings to oppose rental and transport fare hikes became crucial organizing grounds. But infrastructure also became a political terrain in much more mundane and less clearly symbolic and visible ways, from bus and rent boycotts to the nonpayment of service charges, illicit settlement, and multiple small acts of sabotage.[24] Such "popular illegalities" often appeared at seemingly neutral, technical-administrative links to the state.[25] And while the liberation movement at times explicitly politicized such acts, often this was a politics that remained, indeed *had to* remain, on a murkier administrative terrain.

In these diverse ways, infrastructures also became mundane sites for the cultivation of political subjectivities, not merely in the form of ideology, but also in the production of oppositional habits and affective investments as well as embodied stances against the state. Protests or small acts of sabotage were often highly localized affairs, triggered by specific events or mundane complaints, such as a rate hike or an eviction. Often they were direct responses to the everyday violence of apartheid. At other moments, they were reactions to the intransigence of local bureaucrats or the capriciousness of administrative fiat. Over time, the liberation movement subsumed such localized protests, boycotts, and popular illegalities within the larger progressivist telos of liberation, rearticulating them in a nationalist language and ascribing to them an intentionality and historicity that often differed from the residents' own experiential realities and multiple forms of reasoning for engaging in them.[26] Partly because the liberation movement invested such acts with its own specific political and symbolic meanings, in the aftermath of liberation, the new ANC-led government expected that these protests and illegalities would automatically end, given the legitimacy and democratic foundation of the new dispensation and given the ANC's particular modernist understanding of politics.[27] And yet protests and popular illegalities did not end in 1994, but continued or reemerged in a multiplicity of ways. In the absence of an authorizing language, such protests became increasingly resignified as criminal, irrational, or simply self-interested acts. Thus, nonpayers today are no longer "rent boycotters," but "free riders" or "economic saboteurs." They no longer negotiate with government over payment, but are adjudicated in courts of law. This is a liberal-secular politics, that is, a politics in which the political and the administrative are clearly distinguishable, where violence is the prerogative of the state, where the foundations of law are no longer in question.[28]

However, precisely because apartheid worked against the creation of a common public, there is also a long tradition of politics that does not work via the public, that does not assume free circulation, and that has historically often taken shape on an infrastructural terrain. Importantly, this is a politics that is less easily apprehended through normative liberal understandings of politics, in part because it is not by definition a liberal politics. If apartheid infrastructures shaped a politics in nonpublic forms, it is this modality of politics and disagreement that continues in the present, albeit without a language to render it intelligible.

My suggestion is that the *perspective* of infrastructure enables a different purchase both on the histories of the antiapartheid struggle and, in turn, on the contemporary moment in which protest is often rendered unintelligible. Indeed, the persistence of this infrastructural terrain of politics tells a different, less linear story of the end of apartheid. The moment of liberation in the 1990s was often framed in the grammar of a globally circulating post–cold war discourse of "transition" through which South Africa miraculously emerged as what the late Neville Alexander (2003) felicitously called an "ordinary country." And yet, such framings glossed over the intransigence of apartheid that hovered just below talk of "reconciliation," "conflict resolution," or the "rainbow nation." This vantage point often rendered illegible the persistence of this administrative-infrastructural terrain—the affective registers of the antiapartheid struggle that could not simply be erased, the embodied stances of defiance toward the state and the insurgent subjectivities that lived on after the end of apartheid. Infrastructure, and the bureaucratic and insurgent forms to which it has historically been intimately linked in South Africa, may thus provide a perspective to foreground these registers and to make apparent the *work* of *making* liberal democracy, work that is often incomplete and defined by failures as much as successes.[29]

The Technopolitics of Infrastructure after Apartheid

Let me end by briefly elaborating this political terrain in some more detail by looking at the contemporary politics of infrastructure in Soweto. For some years now, I have followed a controversial large-scale project to upgrade infrastructure and install prepaid water meters in all Soweto households. Called

Operation Gcin'amanzi (Zulu for "save water"), the project was initiated in 2003 by the company Johannesburg Water. The utility, though still publicly owned, had been recently corporatized and was managed by a subsidiary of the multinational Suez Group. The project, the largest water prepayment metering installation in the world, ended the apartheid-era unmetered water connections that had given residents an unlimited supply of water, which was now seen as inefficient and environmentally and financially unsustainable. Instead of the flat-rate connections that had served most Sowetans until then, each plot in Soweto was being fitted with a prepaid water meter that automatically dispensed the small nationally mandated free basic amount of water every month, after which residents had to purchase water credits from the local utility offices, in order to avoid automatic self-disconnection. Importantly, the project was decided upon by the recently corporatized water utility with no consultation with affected communities. While Operation Gcin'amanzi was supposed to have been completed in 2009, it is in part due to the resistance it encountered that it is still not complete today.

Over the years, the project became a political terrain in several forms, each of which I want to very briefly map here. First, almost immediately after the project was begun in the Phiri section of Soweto, a residents' group formed and called themselves the Phiri Concerned Residents Forum, affiliated with the Anti-Privatisation Forum. The group immediately began organizing demonstrations. Thus, Operation Gcin'amanzi became the subject of intense protest, with residents digging out pipes that had been laid for it and organizing marches and demonstrations. At times, the meters, ripped out from walls and backyards, were taken to demonstrations and performatively carried in hands and on women's heads.[30] At other times, sewage was spilled outside local government offices. Later, in a countercampaign, which activists termed Operation Vul'amanzi (Zulu for "let the water flow") "guerrilla plumbers" bypassed the meters and reconnected the water supply. During such protests, they would battle the police, and eventually a private security firm policed each construction site in what one utility official likened to a "state of war." Here, infrastructure became a symbolic tool, as meters and pipes were ripped out from backyards and performatively displayed at protests. This was unmistakably a politics of visibility that turned the project and indeed the meters themselves into a "matter of concern."

When the protests died down, in part due to the rising numbers of arrests, five residents took the utility to court on the basis of their constitutionally guaranteed right to water. Here, the project became subject to what Jean and John

Comaroff (2006) have described as "lawfare"—a politics by legal means. As the case wended its way up to the Constitutional Court, it also produced a specific judicial technopolitics. On a legal plane, the residents' multiple concerns about the project were turned into competing expert calculations of "basic needs." Thus, both during the protests and in a different way in the constitutional court case, infrastructures became matters of explicit deliberation and disagreement, objects around which specific publics could be formed. In this sense, although this politics took shape in locations outside of the formal political sphere—the street and the court—this was still a public kind of politics.[31]

There was, however, a third, less public mode of politics, in which the project's infrastructure itself became a political terrain. By 2011, 30–40 percent of the meters had been illicitly bypassed, and engineers struggled to develop and retrofit security features. Thus, it became clear that the battle over payment, which began during the late apartheid period, did not end with the installation of the meters. Rather it was increasingly delegated to a technical terrain and carried out by engineers, embedded in security features and specific forms of expertise. Here, indeed, bypassing and securing the meters produced a specific technopolitics, in which minute technical transformations became tactical moves in an ongoing war over payment, basic needs, and, ultimately, postapartheid citizenship.

There are numerous examples of how this "technical politics" unfolded in Soweto, but let me here focus on the story of one resident, Eunice Ngwenya, who participated in each of these three forms of infrastructural politics: she was an active participant in the residents' group and she also became one of the applicants in the court case.[32] Like many other Soweto residents, Eunice had not paid for her services in a long time. When the meter project was first announced, Eunice, alongside several other Phiri residents, had refused to have the meter installed, fearing she would not be able to pay for water. In response, the utility at first cut resistant residents off from the water supply entirely. For Eunice this meant that she needed to find another way to get water. Her neighbors' initial help soon evaporated when their water supply was threatened, too. Thus, Eunice began walking with buckets to the nearest public tap. Because most public taps had been eliminated after the prepayment project and there were no other water sources available, this required a three-kilometer walk to a part of Soweto that the project had not yet reached. Because she needed a lot of water to care for her niece who was suffering from AIDS, Eunice had to do this walk several times a day.

Eventually, the utility decided to provide residents who had refused a meter with a standpipe in the yard. After six months, Eunice agreed to the installation. While this meant that she no longer needed to walk to get water, she could still not use her kitchen taps or flush her toilet, because the city had "downgraded" her service from a level of service (LOS) 3 (the highest level of public service provision, indicating water-borne sewage) to LOS 2 (only a standpipe in the yard, but no direct water connections to service the internal plumbing).[33] This was an intentionally punitive move that meant that Eunice still could not use her kitchen faucet or her toilet. However, with help from her husband, she constructed an intricate system of hoses that illicitly reconnected at least the toilet. This allowed her to flush the toilet without having to use buckets. She had to do so carefully though, because, if discovered, she would have to pay a R4000 fee to the utility. Importantly, here punishment for objections to the water project was rendered by administrative action, indeed, infrastructurally.

The point of this story is not so much to demonstrate the relative deprivation of Soweto residents. Indeed, as city officials are quick to point out, there are many places in South Africa, especially in informal settlements and rural areas, that still have no formal water connection at all.[34] Rather, this anecdote is significant because it shows the technopolitical forms in which the relationship between the utility and residents is constituted and with which arguments about Operation Gcin'amanzi were ultimately resolved. Indeed, it is this specific form of engagement that often is at the heart of many residents' outrage. Most obvious and often remarked upon by residents is the manner in which the utility approached residents, which many activists argued was "technocratic." Indeed, the High Court judge who first ruled on the project argued that Operation Gcin'amanzi had been defined by an apartheid-style "patronization" of township residents that discriminated on the basis of race and gender.[35]

And yet it is necessary to unpack the "technocratic" further. For the problem was not simply that the utility mindlessly approached this problem as a "technical" one by depoliticizing it. Rather this was a *particular* technopolitics, that is, a politics carried out on an infrastructural terrain in a way that to many residents was eerily familiar. Viewed against the backdrop of the long histories of apartheid technopolitics I have sketched in this chapter, it was continuous with the technical forms that had defined the late apartheid era and the anti-apartheid struggle along with the infrastructural modalities it took during the 1980s. Indeed, here again infrastructure became the material terrain onto which disagreement was delegated. Moreover, for many residents, such technical forms of engagement and intervention re-affirmed the mysteriousness of local government operations in which apportioning blame is not easy. In this sense, the infrastructural continues to be the material embodiment of relations to the state that are often experienced as opaque.

Whether in disputes over payment for service charges, water cut-offs, or evictions, in the spectacular destruction of infrastructures during protests, or in the more silent bypassing of water meters, in Soweto and many other townships around South Africa, infrastructures themselves thus continue to be objects and forms of making political claims. Neither merely symbols of nor simply conduits for power, here infrastructures at times emerge as the very terrain of politics, one that has been historically constituted over decades in a diversity of ways. Administrative links to the state—infrastructures, fiscal relations, and judicial mechanisms—thus become sites at which ethical and political questions that were central to the antiapartheid struggle continue to be mediated, negotiated, and contested, albeit in often nonpublic forms, without an authorizing language that render them intelligible as "political."

This is particularly so in the paradoxical postapartheid conjuncture of liberation and liberalization (cf. Comaroff and Comaroff 2001) in which citizenship has been extended to an unprecedented number of people at the same time as the entitlements conferred by citizenship are often in question. In a context in which normative locations of politics are often de facto inaccessible to many township residents, ethnographically studying infrastructures thus has the potential to attend to a politics that takes shape in unexpected locations and unfamiliar technical forms. Questions concerning citizenship, belonging, or civic virtue may here be expressed by flicking a switch, cutting a wire, or installing a standpipe. This is a politics far removed from the liberal imaginaries of a transparent, unencumbered sphere of public deliberation and circulation. This is a much more encumbered story, of material political claims, of less visible stances of defiance, and of "embodied memories" of a more durable kind.

Conclusion

To conclude, let me briefly return to Malema's toilet revolution and Cape Town's toilet elections. While the spectacular mobilization of toilets turned infrastructure into a matter of concern, the anxious public debates about populist threats to democracy could not quite capture the extent to which this idealized liberal democracy was never quite achieved. In a very real sense, there is today a disjuncture between the miracle of liberal democracy with its free and fair elections, rights discourses, party politics, and civil society, on the one hand, and, on the other, local, lived experiences of democracy that are often at least in part shaped by late apartheid and its infrastructures and associated administrative practices. I have suggested, then, that there is a less visible relationship between infrastructure and democracy, and, in turn, that there is a less public mode of politics involving infrastructure, one that becomes intelligible only against the backdrop of the material histories of the antiapartheid struggle.

Beyond ethnographic object, infrastructure thus also provides an epistemological vantage point, a perspective from which one might think differently about South Africa's "transition." From the point of view of the multiple infrastructural, administrative relations to the state, and thus from an imaginative horizon less constrained by the dominance of juridico-political periodization, an accounting for apartheid's less visible legacies and their material and affective durability becomes possible. Exploring these more subterranean, administrative-technical transformations of governing urban areas also points us to the ways in which governmental and juridico-political transformations are often "out of sync" and shaped by distinct temporalities. From this perspective, while the first general election in 1994 marks the beginning of democracy in South Africa, the continuities and discontinuities with the late apartheid era often unfold at different speeds and in distinct registers.

While the waves of contemporary protests are clearly often bound up with material demands on the state, they *also* often reflect the ways in which apartheid's intransigence is materialized in roads, pipes, bureaucratic forms, administrative fiat, and indeed embodied forms of ethical and political knowledge, an intransigence that is reinforced in the present democratic moment. Retelling this history from the perspective of its infrastructure thus also enables a view of the much longer histories of the disjunctures of urban citizenship in South Africa, and the ways in which the liberal order of free circulation often rests on illiberal foundations. In this sense, this is also decidedly *democracy's* infrastructure.

To return, then, to Hacking's and Scott's challenges, what is the point of an anthropology of infrastructure? Surely, there are many answers to this question as the growing scholarship on infrastructure shows. However, taking seriously Scott's (1999) provocative reflections on criticism, it also becomes apparent that the epistemological *work* that infrastructure can do cannot be guaranteed in advance, but is constituted always in relation to specific conceptual and political problem spaces and historically constituted fields of intervention. In other words, its epistemological promise is at once contingent and strategic. For my purposes here, infrastructure is both an ethnographic object and an epistemological vantage point from which to understand a less apparent post-apartheid political terrain and a location from which the South African present may be defamiliarized and the political rethought.

1 See *Mail and Guardian*, "Malema blasts ANC for Lack of Housing and Toilets," July 27, 2013.

2 *Mail and Guardian*, "The Toilet Election," May 13, 2011.

3 *Mail and Guardian*, "Toilet Election."

4 See here, for example, Ernesto Laclau's analysis of populism and its link to "dangerous excess which puts the clear-cut moulds of a rational community in question" (2005: x). With a more specific focus on South Africa, see Deborah Posel's analysis (2014) of Malema's populism and Jean Comaroff's reflections (2011) on populism's relation to late liberalism.

5 Some of them have been usefully taken up in Brenda Chalfin's (2014) study of what she terms "infrastructure of bare life" in Ghana. For a close discussion of the "poo wars" in Cape Town, see Steven Robins (2014). Both in different ways draw on Arjun Appadurai's discussion (2001) of the "politics of shit."

6 I here in part follow work by scholars like Michael Warner, Beth Povinelli, Charles Hirschkind, and others who have contributed to a retelling of the shape of "publics" and "counterpublics," including its material forms. See also, e.g., Anand (2012); Barry (2013); Fennell (2015); Mitchell (2011). On modularity, see Hannah Appel's analysis (2012) of the "making of modularity" in relation to oil corporations.

7 One might think here also of De Laet and Mol's "Zimbabwe bush pump," which, they argue, "inspires love" (De Laet and Mol 2000; cf. Redfield 2012); Brian Larkin's depiction of the opening of awe-inspiring colonial electricity works as the "colonial sublime" (Larkin 2008); or the affective power of bricks in (post)socialist Vietnam (Schwenkel 2013). See Latour (2004) on "matters of concern."

8 On the materiality of the formal political sphere, see, e.g., Lynch et al. (2005); Barry (2001).

9 I use the term "technopolitics" following Timothy Mitchell's (2002, 2011) and Gabrielle Hecht's (2009) work, but I extend it to encompass not merely large-scale engineering projects, but also a more micrological technopolitics involving habits, subjectivities, and affective relations.

10 Chipkin has since elaborated this argument in several ways (see, e.g., Chipkin and Meny-Gilbert 2011).

11 Much work in anthropology attests to the methodological difficulties of studying the state and its experiential realities (see, e.g., Gupta 2012; Das and Poole 2004).

12 See her reflections on "imperial debris" (Stoler 2013). See also Sharad Chari's incisive reflections on "detritus" in Durban in the same volume(Chari 2013).

13 Indeed, from the perspective of urban citizenship and the everyday experience of the local state, we might periodize the transition quite differently. While much focus was on CODESA, the national negotiations that took place in Kempton Park in the early

1990s, which would ultimately negotiate the compromise that ended apartheid, this period was also defined by numerous individual local government negotiations between the civic movement and local authorities in many municipalities, which often began and were resolved at different times. From the perspective of urban citizenship and the shifts in governing the urban in South Africa one could thus write a history of political transformation that begins nearly forty years earlier with the 1976 Soweto uprising, shifts dramatically throughout the 1980s, and that in important ways is incomplete in the present.

14 See, in particular, von Holdt et al. (2011) for analysis and case studies on such protests.

15 In Soweto, where I carried out much of my fieldwork, such protests were often concerned with the increasing introduction of prepaid water and electricity meters designed to curb the nonpayment of service charges or illicit reconnections that had persisted since the antiapartheid payment boycotts of the 1980s.

16 Jacques Rancière (1999) usefully distinguishes between "voice" and "speech" in his discussion of politics as "disagreement" (*mésentente*).

17 I draw here in part on the large literature on colonial infrastructures (see, e.g., Mitchell 1991; Prakash 1999; Larkin 2008), but it is important to note that apartheid was distinguished from other colonial experiences by its hypermodernity, racialized bureaucracy, and industrial labor conditions, especially from the 1950s and 1960s onward.

18 See Theodore Porter's discussion (1996) of "trust in numbers" and Posel's analysis (2000) of the "mania of measurement" that defined apartheid. See also Keith Breckenridge's analysis (2014) of the long history of biometric infrastructures in South Africa, which complicates previous arguments in relation to the effect of such state knowledge practices.

19 For further literature on links among race, infrastructure, and urban planning during the period of segregation and early apartheid, see Swanson (1977), Parnell and Mabin (1995), and Robinson (1996).

20 See, in particular, recent fictional and autobiographical accounts, e.g., Jacob Dlamini's *Native Nostalgia*, but also the large body of South African urban social history (Mabin 1992; Bonner and Segal 1998; Gaule and Nieftagodien 2012).

21 In the local areas, the apartheid state appeared in multiple forms, ranging from direct military intervention of the security and armed forces to more indirect forms of government, such as the Black Local Authorities instituted in the early 1980s.

22 Yet, as both Jacob Dlamini (2009) and Adam Ashforth (2006) have shown in different ways, township residents often turned such homogeneous spaces into particular places through practices of naming and minor transformations in yards and houses.

23 On the productivity of this "intransigence" of infrastructure, see Collier (2011).

24 See here, e.g., Murray (1987) and Seekings (1988). In his memoir, a former Soweto administrator noted at the time that buses were stoned or burned by residents "who regard the very existence of these buses as part of the 'system,' i.e. as part of the machinery of apartheid oppression" (Grinker 1986: 42). He also noted the "large number of street light poles which have been damaged to the point of no longer being

functional" (50), which, according to Soweto officials, was due to impatience with the electrification scheme.

25 The term "popular illegalities" is used by Michel Foucault in *Discipline and Punish* and taken up productively by Greg Ruiters in relation to South Africa (Ruiters 2007).

26 These divergences emerge particularly clearly in the archives of the civic movements in Soweto, which show, for example, the extent to which the official rent boycott was in fact preceded by isolated incidents of nonpayment that were only later corralled into a larger movement. See, e.g., "SPD Evaluation" (n.d.), Planact Collection 21.5.19.1, South African History Archives, Johannesburg.

27 In relation to payment, for example, the assumption prevailed that "local government structures would hold sufficient moral currency to sway people who had not been paying in the past," which is testified to most starkly by the fact that local budgets were drawn up with the assumption that payment levels would be at 100 percent (Corrigan 1998: 13).

28 See here Comaroff and Comaroff (2006) as well as Rosalind Morris's (2006) account of the resignification of violence and crime in postapartheid South Africa.

29 Indeed, ending apartheid required work at many often extremely local, micropolitical, and symbolic registers. For example, the establishment of liberal democracy after apartheid required a much more piecemeal labor—to construct a collective understanding of civic virtue, to habituate South Africans to a particular relation to the postapartheid state, to reestablish the authority of the rule of law. It required the demobilization of certain political stances and habits that had been central to the antiapartheid struggle (e.g., to delink questions of service provision from political questions of citizenship; to construct an administrative and a political sphere, which had blurred in the context of late apartheid; to delegitimize violence as a form of political expression; and so forth).

30 These were often in fact electricity prepaid meters.

31 I elaborate these arguments elsewhere (von Schnitzler 2016).

32 Following anthropological conventions, I use pseudonyms throughout.

33 These are the city's official categories to indicate "level of service." Activists have contested the categorization of prepaid meters as LOS 3.

34 While Soweto, given its status as a formal township, is often viewed as relatively privileged, this is yet again a determination made on the basis of access to formal infrastructures—pipes, grids, roads. Given unemployment rates, which approach 50–70 percent in many poorer areas of Soweto, and the fact that many of Soweto's residents live in shacks within the backyards of formal houses, the distinction between "formal" and "informal" is more often tenuous at best. With the increasing deployment of prepaid meters this intensified, given in particular that metering is on a plot rather than household basis, de facto excluding the large number of backyard shack dwellers.

35 See *Mazibuko and Others v. City of Johannesburg and Others*, Johannesburg High Court ruling, 2008, para. 153.

REFERENCES

Alexander, Neville. 2003. *An ordinary country: Issues in the transition from apartheid to democracy in South Africa*. Essen: Berghahn.

Anand, Nikhil. 2012. Pressure: The PoliTechnics of water supply in Mumbai. *Cultural Anthropology* 26(4): 542–564.

Appadurai, Arjun. 2001. Deep democracy: Urban governmentality and the horizon of politics. *Environment and Urbanization* 13(2): 23–43.

Appel, Hannah. 2012. Offshore work: Oil, modularity, and the how of capitalism in Equatorial Guinea. *American Ethnologist* 39(4): 692–709.

Ashforth, Adam. 2006. *Witchcraft, violence, and democracy in the new South Africa*. Chicago: University of Chicago Press.

Barry, Andrew. 2001. *Political machines: Governing a technological society*. London: Athlone.

———. 2013. *Material politics: Disputes along the pipeline*. New York: John Wiley and Sons.

Bonner, Philip, and Lauren Segal. 1998. *Soweto: A history*. Cape Town: Maskew Miller Longman.

Bowker, Geoffrey C., and Susan Leigh Star. 2000. *Sorting things out: Classification and its consequences*. Cambridge, MA: MIT Press.

Breckenridge, Keith. 2014. *Biometric state: The global politics of identification and surveillance in South Africa, 1850 to the present*. Cambridge: Cambridge University Press.

Chalfin, Brenda. 2014. Public things, excremental politics, and the infrastructure of bare life in Ghana's city of Tema. *American Ethnologist* 41(1): 92–109.

Chari, Sharad. 2013. Detritus in Durban: Polluted environs and biopolitics of refusal. In *Imperial debris: On ruins and ruination*, ed. Ann Laura Stoler, 131–161. Durham: Duke University Press.

Chaskalson, M., K. Jochelson, and J. Seekings. 1987. Rent boycotts, the state, and the transformation of the urban political economy in South Africa. *Review of African Political Economy* 14(40): 47–64.

Chatterjee, Partha. 2004. *The politics of the governed: Reflections on popular politics in most of the world*. New York: Columbia University Press.

———. 2011. *Lineages of political society: Studies in postcolonial democracy*. New York: Columbia University Press.

Chipkin, Ivor. 1995. Citizenry and local government: A new political subject? *Indicator South Africa* 13: 37–40.

Chipkin, Ivor, and S. Meny-Gilbert. 2011. *Why the past matters: Histories of the public service in South Africa*. Johannesburg: PARI Short Essays.

Clark, Nancy L. 1994. Manufacturing apartheid: State corporations in South Africa. New Haven: Yale University Press.

Collier, Stephen J. 2011. *Post-Soviet social: Neoliberalism, social modernity, biopolitics*. Princeton, NJ: Princeton University Press.

Comaroff, Jean. 2011. Populism and late liberalism: A special affinity? *Annals of the American Academy of Political and Social Science* 637(1): 99–111.

Comaroff, Jean, and John L. Comaroff, eds. 2001. *Millennial capitalism and the culture of neoliberalism*. Durham: Duke University Press.

———. 2006. *Law and disorder in the postcolony*. Chicago: University of Chicago Press.

Corrigan, Terrence. 1998. *Beyond the boycotts: Financing local government in the post-apartheid era*. Johannesburg: South African Institute of Race Relations.

Das, Veena, and Deborah Poole, eds. 2004. *Anthropology in the margins of the state*. Santa Fe, NM: School for Advanced Research (SAR) Press.

De Laet, Marianne, and Annemarie Mol. 2000. The Zimbabwe bush pump: Mechanics of a fluid technology. *Social Studies of Science* 30(2): 225–263.

Dlamini, Jacob. 2009. *Native nostalgia*. Johannesburg: Jacana Media.

Dubow, Saul. 1995. *Scientific racism in modern South Africa*. Cambridge: Cambridge University Press.

Edwards, Paul N., and Gabrielle Hecht. 2010. History and the technopolitics of identity: The case of apartheid South Africa. *Journal of Southern African Studies* 36(3): 619–39.

Evans, Ivan Thomas. 1997. *Bureaucracy and race: Native administration in South Africa*. Berkeley: University of California Press.

Fennell, Catherine. 2015. *Last project standing: Civics and sympathy in post-welfare Chicago*. Minneapolis: University of Minnesota Press.

Foucault, Michel. 1977. *Discipline and punish: The birth of the prison*. New York: Pantheon.

Gaonkar, Dilip Parameshwar, and Elizabeth A. Povinelli. 2003. Technologies of public forms: Circulation, transfiguration, recognition. *Public Culture* 15(3): 385–397.

Gaule, Sally, and Noor Nieftagodien. 2012. *Orlando West, Soweto: An illustrated history*. Johannesburg: Wits University Press.

Grinker, David. 1986. *Inside Soweto: The inside story of the background to the unrest*. Johannesburg: Eastern Enterprises.

Gupta, Akhil. 2012. *Red tape: Bureaucracy, structural violence, and poverty in India*. Durham: Duke University Press.

Hacking, Ian. 1999. *The social construction of what?* Cambridge, MA: Harvard University Press.

Hecht, Gabrielle. 2009. *The radiance of France: Nuclear power and national identity after World War II*. Cambridge, MA: MIT Press.

Hindson, Doug. 1987. *Pass controls and the urban African proletariat in South Africa*. Johannesburg: Ravan Press.

Jolobe, Zwelethu. 2014. The crisis of democratic representation in local government. Paper delivered at the Wits Institute for Social and Economic Research (WISER), University of the Witwatersrand.

Laclau, Ernesto. 2005. *On populist reason*. London: Verso.

Larkin, Brian. 2008. *Signal and noise: Media, infrastructure, and urban culture in Nigeria*. Durham: Duke University Press.

———. 2013. The politics and poetics of infrastructure. *Annual Review of Anthropology* 42(1): 327–343.

Latour, Bruno. 2004. Why has critique run out of steam? From matters of fact to matters of concern. *Critical Inquiry* 30(2): 225–248.

Lynch, Michael, Stephen Hilgartner, and Carin Berkowitz. 2005. Voting machinery, counting and public proofs in the 2000 US presidential election. In *Atmospheres of democracy*, ed. Bruno Latour and Peter Weibl. Cambridge, MA: MIT Press, 814–825.

Mabin, Alan. 1992. Dispossession, exploitation and struggle: An historical overview of South African urbanization. In *The apartheid city and beyond: Urbanization and social change in South Africa*, ed. David M. Smith. London: Routledge, 12–24.

Mbembe, Achille. 2003. Necropolitics. *Public Culture* 15(1): 11–40.

Mitchell, Timothy. 2002. *Rule of experts: Egypt, techno-politics, modernity*. Berkeley: University of California Press.

———. 2011. *Carbon democracy: Political power in the age of oil*. London: Verso.

Morris, Rosalind C. 2006. The mute and the unspeakable: Political subjectivity, violent crime and "the sexual thing" in a South African mining community. In *Law and disorder in the postcolony*, ed. Jean Comaroff and John Comaroff, 57–101. Chicago: University of Chicago Press.

Murray, Martin. 1987. *South Africa: Time of agony, time of destiny: The upsurge of popular protest*. London: Verso.

Parnell, Susan, and Alan Mabin. 1995. Rethinking urban South Africa. *Journal of Southern African Studies* 21(1): 39–61.

Porter, Theodore M. 1996. *Trust in numbers: The pursuit of objectivity in science and public life*. Princeton, NJ: Princeton University Press.

Posel, Deborah. 1991. *The making of apartheid, 1948–1961: Conflict and compromise*. Oxford: Clarendon Press.

———. 2000. A mania for measurement: Statistics and statecraft in apartheid South Africa. In *Science and society in Southern Africa*, ed. Saul Dubow, 116–142. Manchester, UK: Manchester University Press.

———. 2014. Julius Malema and the post-apartheid public sphere. In special issue, Rethinking the publics, of *Acta Academica* 46(1): 32–54.

Prakash, Gyan. 1999. *Another reason: Science and the imagination of modern India*. Princeton, NJ: Princeton University Press.

Rancière, Jacques. 1999. *Disagreement: Politics and philosophy*. Minneapolis: University of Minnesota Press.

Redfield, Peter. 2012. Bioexpectations: Life technologies as humanitarian goods. *Public Culture* 24(66): 157–184.

Robins, Steven. 2014. The 2011 toilet wars in South Africa: Justice and transition between the exceptional and the everyday after apartheid. *Development and Change* 45(3): 479–501.

Robinson, Jennifer. 1996. *The power of apartheid: State, power and space in South African cities*. Oxford: Butterworth-Heinemann.

Ruiters, Greg. 2007. Contradictions in municipal services in contemporary South Africa: Disciplinary commodification and self-disconnections. *Critical Social Policy* 27(4): 487–508.

Saul, John S., and Stephen Gelb. 1981. *The crisis in South Africa: Class defense, class revolution*. New York: Monthly Review Press.

Scott, David. 1999. *Refashioning futures: Criticism after postcoloniality*. Princeton, NJ: Princeton University Press.

———. 2005. The "social construction" of postcolonial studies. In *Postcolonial studies and beyond*, ed. Ania Loomba, 385–400. Durham: Duke University Press.

Schwenkel, Christina. 2013. Post/socialist affect: Ruination and reconstruction of the nation in urban Vietnam. *Cultural Anthropology* 28(2): 252–277.

Seekings, Jeremy. 1988. Political mobilization in the black townships of the Transvaal. In *State, resistance and change in South Africa*, ed. Philip H. Frankel, Noam Pines, and Mark Swilling, 197–228. Johannesburg: Southern Book Publishers.

Stoler, Ann Laura, ed. 2013. *Imperial debris: On ruins and ruination*. Durham: Duke University Press.

Swanson, Maynard W. 1977. The sanitation syndrome: Bubonic plague and urban native policy in the Cape Colony, 1900–1909. *Journal of African History* 18(3): 387–410.

von Holdt, Karl, Malose Langa, Sepetla Molapo, Nomfundi Mogapi, Kindiza Ngubeni, Jacob Dlamini, and Adèle Kirsten. 2011. *The smoke that calls: Insurgent citizenship, collective violence and the struggle for a place in the new South Africa: Eight case studies of community protest and xenophobic violence*. Johannesburg: Centre for the Study of Violence and Reconciliation.

von Schnitzler, Antina. 2016. *Democracy's infrastructure: Techno-politics and protest after apartheid*. Princeton, NJ: Princeton University Press.

Warner, Michael. 2002. Publics and counterpublics. *Public Culture* 14(1): 49–90.

Zuern, Elke. 2011. *The politics of necessity: Community organizing and democracy in South Africa*. Milwaukee: University of Wisconsin Press.

6

A Public Matter: Water, Hydraulics, Biopolitics
NIKHIL ANAND

Mumbai's hydraulic infrastructure has been incrementally extended over the last 150 years, and today it is the sixth largest in the world. More than seven thousand city employees distribute nearly 3.5 billion liters of water daily by managing at least four thousand kilometers of pipe. Their work makes possible the lives of twelve million of the city's residents, generating handsome revenue surpluses for the city. Yet although the ledgers of the city's water department are overflowing with funds, maintenance works contracts, and indeed water, the city's hydraulic network does not distribute water continuously to individuated households for twenty-four hours a day. Instead, water is distributed only to "co-operative housing societies" and on a schedule. Most housing societies—in slums and high-rises alike—receive water for two to three hours a day. They then deploy various technologies to gather this water during the hours of supply and to distribute it among their members.

Between 2003 and 2009, the Municipal Corporation of Greater Mumbai delegated a team of World Bank–appointed management consultants, Castalia Advisors, to reform and "improve" the distribution regime in one ward of the city. During this time, the consultants sought to transform the network from an intermittent system of scheduled supply to one in which water would be available to users 24/7. They attempted to convert the existing system from one in which the quantities of water distribution were rationed through a public water schedule to one in which consumption would be regulated by price (see also Collier 2011; von Schnitzler 2016). In so doing, they proposed to shift the locus of regulation from public state engineers to private water meters charging "rationalized" prices.[1] The consultants at Castalia insisted that, with these changes, water would be more efficiently and equitably distributed. This effort failed spectacularly in 2008 for a variety of reasons, not least because the consultants were unable to stabilize their measures of water during the water audit (Anand 2017). The reform effort also ran into trouble when settlers (also called slum dwellers) opposed the pilot project, both through their daily practices and by insisting that water was and should remain a public good.

In this chapter, I draw attention to the ways in which marginalized residents regularly produce water as a public matter in the offices of city councilors and public hydraulic engineers. If biopolitics is the art of managing populations, as Michel Foucault has pointed out, an attention to these demanding practices describes how hydraulic publics are made and managed in the city. In Mumbai, water services are not extended (or pushed) in to the settlements, by the seeing knowing state (Scott 1998), eager to make its subjects through the extension of live making services. Water services are sites of significant activity, called for by residents eager to make life material and meaningful in the hydraulic city. Having become accustomed to access water from public agencies through material claims, settlers are very uneasy by the prospect of privatization of the city's public water-distribution system.

In recent years, anthropologists have critically engaged with the work of Jürgen Habermas and Michael Warner to critically interrogate the ways in which publics are made through the circulation of texts and the ways in which these are represented, reproduced, and narrated (Habermas 1989; Warner 2002; Cody 2011). If documents, plans, and policies are important sites where publics are imagined, they are *materialized* through a series of social *practices,* such as the extension, management, and maintenance of water infrastructures. Carrying the life needs for residents and businesses of particular neighborhoods, water pipes constitute collectives that are joined by their experience of water supplies. Hydraulic publics do not always belong to the same socioeconomic class (Björkman 2015). When very wealthy and very poor residents live in neighborhoods served by the same water line, as they often do in Mumbai, they constitute publics that are often surprisingly heterogeneous.

I demonstrate how water services often become a matter of public concern, even as the municipal government works toward ensuring that water remains a private matter.[2] Residents frequently take their concerns about water—a domestic matter—out of the privacy of their homes and bring it up in the offices of councilors and administrators. They gather as collectives and demand and call for state officials in very public ways to attend to their water problems. They also demand more water by making claims on the city's hardware: its pipes, meters, and pumps. Taken together, residents frequently deploy different practices to call out for water and to make it a public matter.

Before I go further, first a brief note to explain what I mean when I invoke the term "public." Jeff Weintraub and Krishan Kumar (1997) have suggested we consider four different usages of the public in social theory and its distinctions from the private sphere. The "public," according to them, is used as a term in social theory and politics to identify (1) public *goods* (juxtaposed against those distributed by the market economy), (2) as a political *community* distinct from markets and the state (as the Habermasian public sphere, or Warner's counterpublics), (3) as a type of *social relationship* between strangers that mediates between the bureaucratic public realm and the space of the home (like Jane Jacobs's eyes on the street), and finally as (4) a gendered category that demarcates a split between the family/ household/*oikos* (as private) and *a larger political order* that is public (Arendt, drawing on Aristotle).

These categories are helpful in noting the appearances of public water in Mumbai. In Mumbai, water is a public good distributed by a public agency—the city's water-supply department. As news about its shortages regularly circulates in the city's newspapers, the city's water is also productive of a political community—a municipal public sphere that is constituted both by anxieties of scarcity and by fears of outsiders and immigrants. As I demonstrate elsewhere, these processes are deeply gendered (Anand 2017). Here, I am particularly interested in kinds of semipublic spaces in which these claims are made and generated. As subjects make claims to water in the public-private offices of city councilors, they blur boundaries between domestic spaces and the political order that constitutes them. These iterative processes of demanding water in the city make and maintain the city and its public officials (see also Gupta 1995). How might we theorize these claims as claims of publics?

In drawing attention to these material practices, I make three related points about Mumbai's hydraulic publics in this chapter. First, Mumbai's first hydraulic publics not only preceded the hydraulic state, but also gave the postcolonial state its *raison d'être* and peculiar institutional form in succeeding decades. In the first section of the chapter, I draw attention to the contingent history on which Bombay's water-supply regime was founded. I do this not so much to provide a historical backdrop. Instead, as the city's water department was formed amid extant public claims to water, in this section I demonstrate how the city's municipal water department was formed as a political project to redirect the city's extant publics to the partly liberal saviors of the colonial state.

Second, I attend to the ways in which marginalized residents living in Mumbai today gather as domestic publics to make claims to water services by drawing on the help of city and state legislators. As residents leave their homes to make claims to water (for their homes) in the offices of city councilors, their practices reinscribe and reinstate them as a "domestic public"—a gendered collective that is made through discrete appeals to the care of the city councilor and other "public" servants in the city.

Finally, I conclude by drawing attention to the ways in which the making of publics and public authorities are subtended by the materialities that compose the city's water. As residents are made public by the manipulation of water pressures, it is difficult for engineers to recognize publics in one location without drawing water away from others. Publics need constant matter, labor, and recognition to be maintained as such.

By focusing on the ways in which water is generative of its publics, I follow recent approaches in anthropology, geography, history, and science and technology studies that have urged we consider the ways in which materialities and technologies call out certain political forms (Braun and Whatmore 2011; Ingold 2012; Larkin, this volume).[3] Drawing on this work, I suggest that both publics and their states are brought into being *with* the discrete, partial, and compromised pipes and liquid materials of water infrastructures that form the city. As "communities of the affected" (Marres 2012), hydraulic publics come into being through the material and intimate political commitments to care for the extensive and enduring consequences of water distribution in Mumbai.

Public Works

Historians have described how modern infrastructures emerged in the middle of the nineteenth century as a key modality for consolidating the project of liberal government. The expansion of roads, pipes, and sewers in the nineteenth century gave the liberal state a material form. The historian Patrick Joyce (2003) describes how, in the proliferation of these infrastructures—in both the colony and the metropole—liberal theorists sought to "free" citizen subjects from the entailments of fragmentary political communities and to instead subject them to the singular authority of the modern state.

This project was critical in colonial Mumbai. As late as the middle of the nineteenth century, residents of the burgeoning city continued to depend on the authority and munificence of wealthy philanthropists and community leaders for the provisioning of many life needs, including water. With residents of the city dependent on a patchwork of caste, kinship, political, and patronage relations to live in the city, colonial administrators wrestled with the diverse locations of sovereignty in the colonial city. In her historical work on Bombay's politics from the sixteenth to the eighteenth centuries, the historian Mariam Dossal (2010) describes how colonial agents worked hard to centralize and harmonize different land-tenure records, judicial functions, and economic activities under a single (and colonial) regime of state and government through a series of different projects.

Water was one of these arenas for governmental action. Until the middle of the nineteenth century, residents were hydrated and lived through a series of fresh-water wells and tanks managed by native philanthropists. Nevertheless, as the city's population grew rapidly, the doubling of the city's population, from 300,000 in 1820 to 644,400 in 1872, placed an extreme strain on the city's existing tanks and wells (Dossal 2010; Gandy 2014). Yet, uncertain of the future (and their future) in the metropolis, colonial administrators were reluctant to invest in the production and management of water infrastructure for the city's publics. They debated as to whether either the city or its citizens were deserving of modern water infrastructures.

A series of world historical events, however, produced a historical conjuncture that strengthened the case for a public dam-driven water-supply system in the second half of the twentieth century. First, following the depredations of the East India Company, a set of exploitations that produced the Revolt of 1857, the British Crown took over the direct administration of the colonial government in Mumbai and began to see its rule over the subcontinent in more direct and enduring ways. Second, severe droughts in 1854 and 1855 exacerbated the precarious state of water services in the city. Livestock were prohibited in the city, and water was brought in on boats and trains to meet the city's water needs.

Finally, two events made Mumbai an extremely profitable outpost of the British empire and generated the funding necessary to produce such a system. The colonial fortunes of the city grew as several Indian and British merchants profited handsomely from the opium trade and the Opium Wars in China that took place between 1839 and 1860 (Farooqui 2006). The fortunes of the city's cotton merchants also grew rapidly in this period, after the opening of the Suez Canal and the beginning of the American Civil War (during which time cotton exports ceased from the United States). Thus, in 1860, when the British government finally decided to construct a large, modern water infrastructure system in British India, it did so with the fiscal resources of the city's colonial administration. It produced a political solution for the city that centralized power in the hands of colonial municipal administrators and engineers.

When the Vihar Dam was completed in 1860, it was foundational to not only the city water department, but also the city's municipal state. The newly engineered, dam-dependent public water system established by the colonial government centralized water provision in the city. As publics went from congealing around wells and tanks, to congealing around pipes, municipal water services marked a new technological and political moment for the city. Indeed, from the very beginning, the city's pipes were generative of its municipal institutions.

FIGURE 6.1 Colonial hydraulics: Nineteenth-century water pipes continue to service the city. Photo by the author.

The Bombay Municipal Corporation was founded in 1882 to finance, manage, and maintain the pipes, dams, and distribution systems of the newly installed public water system (Dossal 1991). Taken together, the beginnings of the municipal water-supply network served to institute the colonial state as the leading institution of water delivery in the city, shift the cost of water delivery onto residents, and rearranged hydraulic publics from those that would form around the tanks, wells, and springs of philanthropists, to those that, over time, would adhere to the life of the water pipes managed and maintained by the colonial city administration (see figure 6.1).

Since the middle of the nineteenth century, the provision of water services has remained key to the political legitimacy of the city's rulers (see Wade 1982) and also has been a critical site at which the shifting geography of the city can be tracked and observed through different moments of expansion. From its modest beginnings—of provisioning thirty-two million liters of water a day through city pipes in 1860—the city's publicly administered water-supply department today provides over 3,400 million liters per day, making it the largest single source of water provision in the city and the sixth largest urban water system in the world. At the same time, the dam-driven water system is not the only water source in the city. While philanthropists seldom gather moral or political authority around matters of well management in the ways they did in the nineteenth century, more than ten thousand wells continue to gather and produce water (and water's publics) in the city. The continued persistence of these wells indicates the persistence of diverse institutional forms in the city.[4]

Today, most residents receive and live on treated water distributed by the Mumbai Municipal Corporation. The water infrastructure is governed by a public system that reproduces difference. By managing times, pressures, and durations of supply, engineers endeavor to allocate 150 liters per person living in authorized buildings. They seek to distribute only sixty to seventy liters per person living in the city's recognized settlements, that is, settlements that have existed prior to 1995 and are deemed through diverse political processes to be deserving of governmental services. Meanwhile, the million and a half residents that live in unrecognized settlements have to make their own arrangements— either by buying water from "recognized" neighbors, or by making surreptitious connections to water mains.

As pipes reticulate from the treatment plans to secondary and then tertiary lines, they divide, constitute, and connect discrete technocratic experts, politicians, and their publics that live and work in the city. In making distribution decisions discreet and invisible, the technocratic designs of experts seek to move a collective good to discrete and private domains (von Schnitzler, this volume).

Yet far from rendering water an administrative, or private, matter, Mumbai's pipes do not preclude the formation of publics. In the next section, I show how piped water is claimed by marginal residents living in a settlement in northern Mumbai. Residents protest, petition, and pressure political leaders to bring reliable access of piped water to their neighborhoods. Settlers and councilors actively work to extend the network. To receive water again, settlers mobilize demands for more water, as publics.

Making Connections

Publics are regularly reproduced through the quotidian maintenance and repair of water infrastructures in Mumbai. When water connections go dry, hydraulic publics today gather in the offices of city councilors (and not those of hydraulic engineers) to demand their share of water. Biopolitics, in this rendering, is not extended from the center as much is it pulled, tugged, and demanded, often quite materially, from the margins. Because city councilors in Mumbai owe their office primarily to the votes of those who live in the city's settlements, they are extensively focused on redressing the discrete grievances of their constituents. Many of them keep their offices open to the public in the evening, when residents of the many settlements they govern can come to have their problems redressed, or at least heard. Residents visit councilor offices with different kinds of problems, including domestic disputes, school admission issues, or health matters. Yet, as councilors would often point out, most of their constituents' problems have to do with restoring water supply to their homes.

In the course of conducting fieldwork in 2007 and 2008, I documented several instances of such complaints in councilors' offices. When water pressure dropped in their service lines, residents generally first approached social workers to arrange a meeting with the city councilor. Because water connections were shared, and because the councilors often responded to their claims more expeditiously, women ensured they went to the councilor's office as a "group." "When we go in a group, then the police get scared," an older woman had once told me when describing how she and her neighbors would pressure state officials to respond. When they go one at a time to solve problems, residents often reported that they were made to wait for long periods, and then sent back without any action being taken. Going in a group, a crowd, a public, I heard, time and time again, was an important means of getting a fair hearing. This gathering alternately entreated, demanded, and shouted at councilors to make the water reappear, as it should, for the "public" (using the English word). They made claims to water based on their kinship or friendship with those favored by the councilor. Councilors frequently noted their complaints and relayed them to the hydraulic engineer in charge of their ward, often at their next visit. Hydraulic engineers, recognizing that the approval of their works contracts was contingent on the good graces of city councilors, frequently found ways to solve these problems. Like councilors, engineers, too, recognized that their failure to resolve the problem would only make the complaints more vociferous. The city's publics, after all, needed water not just to live and to vote, but also to allow engineers to do their work without disruption.

To understand the processes through which settlers have come to access water services in the city, I frequently met with community groups in the settlements where I worked. I first met with Sunita and Rajni tai when they had come to meet Vishnu, their neighbor and the head of Asha, one of the many community centers in the settlement. Sunita and Rajni tai lived in a settlement that was legally entitled to receive pressurized water. Yet, their pipes no longer delivered any water. As they spoke with Vishnu about the problem, I was amazed by the efficacy and familiarity with which each knew what he or she had to do. They collectively wrote a letter on the Asha letterhead and called one

FIGURE 6.2 Service connection, Mumbai. Photo by the author.

of the councilor's workers to set up a meeting. This was the first step, I learned later, in the process to get a new water connection.

When I met Sunita and Rajni at their community center some months later, I was pleased to learn that though it took some time, the councilor had eventually agreed to put a new line in the settlement. Armed with Vishnu's letter, they had met the city councilor to seek his help in solving their water difficulties. In recounting the meeting to me, Rajni tai said, "We told him, we don't want money. Our vote is worth one lakh rupees [US$2,000]. We only want to use that amount now. Because we don't have money. So he put the pipe [in]."

I was slightly surprised about the ways in which Rajni spoke of her meeting with the councilor. Rajni did not consider the meeting as a place where, as a member of civil society, she could speak of water as a right. But neither did she speak of meeting the councilor as a supplicant, as a client requesting exceptional treatment (or a member of political society). Whether she was as explicit at the meeting as she was with me in our interview, I cannot be sure. But through our conversation, she understood that the councilor was obliged to give them a new line. The councilor, she explained, depended on their votes for his position. Not only did they want his permission to lay the water line, they expected him to fund its procurement and installation as well.

Here Rajni articulates a rather mundane and transactional understanding of citizenship, framed around a claim to water. In so doing, she reminds us how the problematization of government and the programs of biopolitics are articulated not just by rulers, but also by the governed. In the weeks following, the councilor did respond. He laid the line, but not before he diverted some of the resulting water to other constituents. In Mumbai, city councilors recognize that their legitimacy is based on their ability to deliver water services. "Water, gutter, passage," councilors would tell me, again and again, when I asked them of their responsibilities and work. During elections, city councilors promise, again and again, improved water services in the event they get elected. Settlers demand and recall these promises once the elections are over. The way in which settlers draw attention to the councilor's moral obligations, to hold up their side of the exchange, shows how it is not only subjects that are responsibilized though the languages of rights and justice (Rose 1999), but also the leaders, dadas (big men), and engineers of the city's public system. As such, the hydraulic public is not only an anonymous population of undifferentiated rights-bearing citizens, nor is it a population that is governed solely by humanitarian recognition and exception. As settlers work to be counted among those deserving of legal water services by reminding councilors and engineers of their promises, the hydraulic public is constituted through languages that hail different notions of justice and help, duty and obligation. It is constituted through a set of intimately known and negotiated relations between settlers, social workers, councilors, and the pipes that connect them.

Such political claims, made through the registers of patronage, kinship, and citizenship, are politically potent in "most of the world" (Chatterjee 2004). In his work on distributive claims in southern Africa, James Ferguson has suggested that such claims are, at their base, political claims for distributive justice that rights advocates and their theorists need to take more seriously (see also Scott 1977):

> For those seeking to build a culture of rights, this apparent yearning for parental authority could only appear a dangerous reversion to the logic of the bad old days of colonialism and apartheid. South Africans, in this view, ought to be equal citizens proudly claiming rights, not child-like dependents seeking the protection of a parent. But in styling the people as "like children," the statement can be read as making a very strong assertion not just of inequality, but also of a social obligation linking state and citizen. In that sense, indeed, sentiments of this kind can provide a basis for powerful political claims based on an obligation so fundamental that it precedes any right.... After all, southern Africans ... have always known that kinship relations such as descent are precisely political rela-

FIGURE 6.3 Cluster of pipes (pipe publics). Photo by the author.

tions, just as anthropologists have always known that such relations are central to structures of distribution in every society. (Ferguson 2013: 236)

In Mumbai, kinship and patronage claims are not only articulated with those of citizenship. The way in which settlers draw attention to the obligations of the city councilor and his workers shows how it is not only subjects that are responsibilized though the languages of rights and kinship, but also the leaders, dadas, and engineers of the city's public system.

While Rajni and Sunita's accomplishments were significant, they were very aware that this achievement—manifest in the arrival of a water pipe—was temporary, fickle, and reversible. In Mumbai, water services are in a state of constant flux. Annual hydrological cycles, main line leakages, shifting demography, the pressures of diverse constituencies, and unanticipated cluster developments constantly compel city engineers to tinker with the water system. As a result, engineers are always changing times, pressures, and duration of supply to keep the system working. Nevertheless, pipes cannot connect all publics, and even when they are extended, they are insufficient to deliver pressured water. Owing to the finite matter of distribution, public services here cannot be extended continuously throughout the city to deserving (or demanding) populations without withdrawing them from others. In the final section of the chapter, I demonstrate how, as water pressure is inversely related to the quantity and quality of pipe materials in the city, engineers need to negotiate, in their everyday work, water's finitude, the redundancy of pipes, and the restive publics that clamor for and need water to live.

Both for legal and material reasons, settlers are particularly subject to the vagaries of changing water pressure in the city. They are permitted to connect to the system only with thin pipes that are especially prone to rupture; also, councilors frequently divert water to other constituents, so settlers like Rajni and Sunita tai frequently find their connections periodically go dry. As settlers demand pressured water on time, and for more time, their demands for a rightful share do not just appear in forms, complaints, or petitions in the offices of politicians and technocrats. They also congeal on the pipes and valves of their localities.

The Materiality of Pressure

The councilor of a different settlement was only too aware of the difficulty his constituents had in accessing water. Like his constituents, the councilor, living in the same settlement, also had the same water difficulties. Some residents would suggest the neighborhood's problems had to do with the malicious labor of *chaviwallas* (key men). They often arrived late and did not open the valves enough, they often told me. Others would suggest it was because of a leaky water main that had remained unrepaired for years.

With water difficulties endemic to the life of the settlement, the councilor worked consistently with the city water department to upgrade and repair existing lines. Crucially, while this labor of pipe repair was necessary, it was not sufficient. You cannot get water simply by purchasing, maintaining, and repairing pipelines, nor do you get it by qualifying for water services through the norms and forms of state regulation. I want to suggest that while formal citizenship—much in the manner of newly laid water pipes—may be additive and accumulative, substantive citizenship, here manifest in pressured water being directed through state pipes to deserving subjects, is generally negotiated and distributed. It needs constant material and social pressure to be maintained (Anand 2011). Thus, while the councilor had managed to direct the repair of city water pipes, he also needed to get the city water department to tinker with the existing network, and to divert water from other neighborhoods to his own, through new arrangements of water valves and the chaviwallas that operated them.

Two years after he took office, the councilor managed to effect a small change. The city water department installed a new valve downstream of Premnagar and began to close it for one hour every day. Through this operation, water backed up into the homes of Premnagar, and its residents received pressured water between 3:30 and 4:30 PM daily. Sure enough, this arrangement soon drew the attention of residents downstream of the line, who began to receive less water. Over the course of a week, they regularly congregated around the valve, literally preventing the chaviwallas from closing the valve at the prescribed time. An altercation and a sit-in ensued. Caught between the needs of Premnagar's residents and those who lived downstream, the engineers were compelled to negotiate a compromise. They agreed to close the valve not for an hour, but for only thirty minutes a day. The negotiated settlement was a poignant reminder not only of the ways in which the city department is not the only authority in control of the city's pipes, but also of the ways in which the city's water pipes and valves constantly and regularly become sites for biopolitical claims, made by differentiated urban populations.

Conclusion

As a number of authors have suggested in this volume (Appel et al.; Harvey; von Schnitzler; Larkin), infrastructures are promising sites to examine the imbrications of aesthetics, political rationalities, meaning, and materiality in everyday life. I hope to have demonstrated how Mumbai's water pipes are generative locations to examine the ongoing relations between states and publics, ideologies and technics. To think with Larkin's generative exposition on form, hydraulic technology and the human emerge in Mumbai through interactions with an enabling milieu. When engineers bring pipes into being, and yet cannot fully control the water that courses through them, their practices indicate a world marked by material excesses that escape both human control and representation.

Despite this excess, the daily practices of chaviwallas, engineers, and councilors nevertheless do materialize connections between the laws, policies, and plans of the polis with the intimate regimes of provisioning in the household: of when water arrives, how long it arrives for, and who needs to collect it. The political contestations around their operation reveal how biopolitical government in Mumbai takes place not only at a distance, nor through self-governing individuals, but through politically potent gatherings around the matter of water distribution in the city. When residents protest on and around water infrastructures for more water, they demonstrate how sovereignty and authority are diffuse forms of power that are negotiated both in the blurred boundaries between state and the home, and also on the materials, pipes, and valves of the water infrastructures that join them. They remind us that water infrastructures are often difficult to govern "at a distance." These ways of accessing water demonstrate how biopolitical regimes around water are at least as tugged and pulled into settlements by their residents as they are extended from the center of states.

Water constantly becomes a source and matter for the making and materialization of promises and publics in various ways. Residents in Mumbai's settlements correctly recognize in mundane water pipes the promise of a durable form of urban belonging at some point in the future. To mobilize pipes and the kinds of life they promise, they recall and call out the election promises that city councilors regularly make for better water services. Holding up electoral promises of times past, they demand and negotiate for a public infrastructure that promises to make them hydraulic citizens in the future. Through these everyday practices of demanding water, the vital material is rendered public, first as a "good" distributed by the state, and, second, as a political matter that gathers and constitutes public bodies in the city. As gendered and classed bodies come out of their homes to work with city councilors, water engineers, and a variety of other strangers in the city, their work shows how water is made and remade as a public matter: one that is not just necessary for life, but also calls out for public institutions to be responsible for its everyday management.

As in the middle of the nineteenth century, so, too, at the start of the twenty-first. When water infrastructures are stretched into the settlement, they materially extend and constitute the city's hydraulic state. As such, the state does not just exist prior to water infrastructures in the city. It is also brought into being—as pipes, regulations, and a distributor of public goods—*through* projects to extend water connections into settlements. The hydraulic state is extended through and because of strong *demands* for a particular kind of modern water, a particular kind of grid life that is demanded by settlers in Mumbai. This water is made a public good, one that is extended and provisioned by a state authority, an authority that distributes differentiated quantities of water to the city's residents.

The second way in which water appears as public matter in the city has to do with the collectives it gathers beyond the private domain in and as more known publics. Residents do not quietly deal with and provision insufficient and unequal supplies of water at home, in private. In one case, women living in Jogeshwari gather and collectively work to resolve their water problems. By working with councilors, social workers, and plumbers, residents constitute publics in their own right, by working beyond the domestic sphere through and with forms of stranger sociality that are very accessible in the city (see

Hansen and Verkaik 2009). Residents in Mumbai constitute themselves as a domestic public, one that is brought into being by neighbors, friends, and social workers outside the home, as they gather around pipes and councilors to solve their water problems. This public is *not* as expansive and deterritorialized as the counter/publics described by Michael Warner (2002), nor is it a public that is only organized and mobilized against the state and corporate capital (Habermas 1989). Instead, this public appears as a "community of the affected," a public that is constituted, collected, and gathered by the shared consequences of water distribution in the city. It is a public that, in Noortje Marres's framing, continues to be constituted as a problem of relevance (Marres 2012). As residents work to restore water pressure to their homes, they do not organize against the state as much as they work to hail the state to distribute the matter of water to their homes (cf. Althusser 1971).

Thus, having learned how to approach the city's authorities with their water problems—through collective, gendered petitions in the offices of councilors—residents of Mumbai's settlements were understandably anxious about projects to "improve" (read: privatize) water-distribution projects in the city. City publics, particularly those living in the settlements, have established a predictable and knowable (if also discretionary) practice of claiming and demanding water in the city through discrete claims in the offices of councilors and engineers. Throughout the improvement project, consultants saw these quiet claims and discrete flows as leakage, a pathology to eradicate through liberal reforms that would "free" the system from interference by councilors and engineers (and their politics). Yet, in contrast, residents in settlements wondered aloud about how and to whom they might register complaints in the future if the city's diverse water authorities were made redundant. I argue that part of the reason the improvement project failed to gather the support of the very residents for whom they claimed to be working was because the proponents of privatization failed to recognize the vitality of water's public forms in a differentiated city. As a result, Mumbai's water network today continues to generate known publics—differentiated communities of care—that demand the return of water amid (and from) mundane, ordinary disruptions in everyday life.

Nevertheless, pipes cannot connect all publics, and even when they are extended, they are insufficient to deliver pressured water. Owing to the finite matter of distribution, public services here cannot be extended continuously throughout the city to deserving (or demanding) populations without withdrawing them from others. Water pressure is inversely related to the quantity and quality of pipe materials in the city, and, as such, engineers need to negotiate, in their everyday work, both water's finitude, the pipe's redundancy, and the restive publics that clamor for and need water to live. As publics are brought into being through the situated materialities and designs of the hydraulic network, these historical formations reveal both the vitality of water and the work of public officials in being responsible for its flows. Composed of officials and pipes, of drinkers and tanks, these more-than-human arrangements of the hydraulic network materialize enduring if unstable forms whose politics continue to matter after they have been constituted (von Schnitzler 2016).

The new pipes, renewed pressures, and compromised timings that authorities produce all recognize some settlers and other residents as deserving citizens of the city. They also produce the state as a social and material actor in the city. Once they are installed, pipes and their associated documents—bills, complaints, and petitions—are infrastructures of commitment. They become a public matter for biopolitical claims and distributional justice, revealing in all their obduracy. They are a resilient, if fickle and familial process through which settlers come to belong to the city and claim a "rightful," if unequal, share of its resources.

NOTES

1 The neoliberal reform of state services has, in recent years, stimulated debate, discussion, and controversy in many parts of the world. Is water more fairly and more inclusively distributed as a public right of citizenship or as a private commodity? Projects to privatize urban water distribution have stimulated widespread protests in many parts of the world, including the Philippines, South Africa, India, and Uruguay. Debates about privatization and neoliberal reform produce critical questions not only about the role of the state, but also about public goods, their efficient and equitable regimes of distribution, and the role of the market. In these debates, the public and the private are categories that are used and theorized in a multiplicity of ways. Indeed, water is, and frequently becomes, a public matter. But how is it public? What *kind* of public and publicity is produced out of the circulation of water's meaning, materiality, and distribution infrastructure? I focus on the ways in which water infrastructures call out for and produce public life.

2 For instance, take the mundane operations of the customer water meter. The water meter is a biopolitical device where the public services of water distribution "meet" its private responsibilities (von Schnitzler 2016). As such, the water meter is a poltical technology that "governmentaliz[es] the state" (Foucault 1991) at the same time as it marks the boundary between the public and the private (Habermas 1989). Therefore, the ways in which the water infrastructure silently distributes and differentiates residents and their water consumption into nuclear households, we might argue, are designed to preclude the formation of publics and instead produce and diffuse a multiplicity of private interests who receive water at home, in private (von Schnitzler, this volume).

3 In an inspiring and provocative collection of essays, Bruce Braun and Sarah What-more remind us that "technicity.... is not merely a supplement to human life. Rather it is originary... The human [has always] come into being *with* this world" (2011: xvii–xviii).

4 Nevertheless, residents who lived where I worked preferred treated municipal sup-plies, in part because they believe it to be a cleaner and more modern source of water (Anand 2012). Marked with the languages of improvement, purity, consistency, and clarity, Mumbai's dam-driven, piped water infrastructure is a source of desire and ac-complishment in the city, and it marks important ways in which hydraulic citizens are recognized as such (Anand 2011).

REFERENCES

Althusser, Louis. 1971. Ideology and ideological state apparatuses (notes towards an in-vestigation). In *Lenin and philosophy and other essays*. New York: Monthly Review Press.

Anand, Nikhil. 2011. Pressure: The PoliTechnics of water supply in Mumbai. *Cultural Anthropology* 26(4): 542–564.

———. 2017. *Hydraulic city: Water and the infrastructures of citizenship in Mumbai*. Dur-ham: Duke University Press.

Arendt, Hannah. 1998. *The human condition*. Chicago: University of Chicago Press.

Aretxaga, Begona. 2003. Maddening states. *Annual Review of Anthropology* 32(1): 393–410.

Bakker, Karen J. 2010. *Privatizing water: Governance failure and the world's urban water crisis*. Ithaca, NY: Cornell University Press.

Barry, Andrew. 2013. *Material politics: Disputes along the pipeline*. New York: Wiley-Blackwell.

Bennett, Jane. 2010. *Vibrant matter: A political ecology of things*. Durham: Duke Univer-sity Press.

Björkman, Lisa. 2015. *Pipe politics, contested waters: Embedded infrastructures of millen-nial Mumbai*. Durham: Duke University Press.

Braun, Bruce, and Sarah Whatmore. 2011. *Political matter: Technoscience, democracy, and public life*. Minneapolis: University of Minnesota Press.

Chatterjee, Partha. 2004. *The politics of the governed: Reflections on popular politics in most of the world*. New York: Columbia University Press.

———. 2011. *Lineages of political society: Studies in postcolonial democracy*. New York: Columbia University Press.

Cody, Francis. 2011. Publics and politics. *Annual Review of Anthropology* 40: 37–52.

Collier, Stephen J. 2011. *Post-Soviet social: neoliberalism, social modernity, biopolitics*. Princeton NJ: Princeton University Press.

Dossal, Mariam. 1991. *Imperial designs and Indian realities: The planning of Bombay City, 1845–1875*. New York: Oxford University Press.

———. 2010. *Theatre of conflict, city of hope: Mumbai, 1660 to present times*. New York: Oxford University Press.

Farooqui, Amar. 2006. *Opium city: the making of early Victorian Bombay*. Gurgaon: Three Essays Collective.

Fennell, Catherine. 2012. The museum of resilience: Raising a sympathetic public in post-welfare Chicago. *Cultural Anthropology* 27(4): 641–666.

Ferguson, James. 2013. Declarations of dependence: labour, personhood, and welfare in southern Africa. *Journal of the Royal Anthropological Institute* 19 (2):223–242.

Foucault, Michel. 1991. Governmentality. In *The Foucault effect: Studies in governmentality*, ed. Graham Burchell, Colin Gordon, and Peter Miller, 87–104. Chicago: University of Chicago Press.

Gandy, Matthew. 2014. *The fabric of space: Water, modernity, and the urban imagination*. Cambridge, MA: MIT Press.

Gupta, Akhil. 1995. Blurred boundaries—the discourse of corruption, the culture of politics, and the imagined state. *American Ethnologist* 22(2): 375–402.

———. 2012. *Red tape: Bureaucracy, structural violence, and poverty in India*. Durham: Duke University Press.

Habermas, Jürgen. 1989. *The structural transformation of the public sphere: An inquiry into a category of bourgeois society*. Cambridge, MA: MIT Press.

Hansen, Thomas Blom, and Oskar Verkaaik. 2009. Introduction: Urban charisma: On everyday mythologies in the city. *Critique of Anthropology* 29 (1): 5–26.

Ingold, Tim. 2012. Toward an ecology of materials. *Annual Review of Anthropology* 41(1): 427–442.

Marres, Noortje. 2012. *Material participation: Technology, the environment and everyday publics*. New York: Palgrave Macmillan.

Joyce, Patrick. 2003. *The rule of freedom: Liberalism and the modern city*. New York: Verso.

Rose, Nikolas. 1999. *Powers of freedom: Reframing political thought*. Cambridge: Cambridge University Press.

Rose, Nikolas, and Peter Miller. 1992. Political power beyond the state—problematics of government. *British Journal of Sociology* 43(2): 173–205.

Scott, James. 1977. Patron-client politics and political change in South East Asia. In *Friends, followers, and factions: A reader in political clientelism*, ed. Steffen W. Schmidt, James C. Scott, Carl Landé, and Laura Guasti, 123–146. Berkeley, CA: University of California Press.

Scott, James. 1998. *Seeing like a state: How certain schemes to improve the human condition have failed*. New Haven, CT: Yale University Press.

Star, Susan Leigh, and James Griesemer. 1989. Institutional ecology, translations and boundary objects: Amateurs and professionals in Berkeley's Museum of Vertebrate Zoology, 1907–39. *Social Studies of Science* 19(3): 387–420.

von Schnitzler, Antina. 2016. *Democracy's infrastructure*. Princeton, NJ: Princeton University Press.

Wade, Robert. 1982. The system of administrative and political corruption: Canal irrigation in South India. *Journal of Development Studies* 18(3): 287–328.

Warner, Michael. 2002. Publics and counterpublics. *Public Culture* 14(1): 49–90.

Weintraub, Jeff Alan, and Krishan Kumar. 1997. *Public and private in thought and practice: Perspectives on a grand dichotomy*. Chicago: University of Chicago Press.

PART III. *Promise*

7

Promising Forms: The Political Aesthetics of Infrastructure

BRIAN LARKIN

Infrastructures, as technical objects, take on form. Once something exists—
say a road or an electric plant—we are not just in the domain of matter but of
technological ensembles that are enformed as they are brought into material
existence. In the study of infrastructures, form is both ubiquitously visible yet
absent from analytic consideration. However, it is the interface through which
humans engage with technologies and is part of the reciprocal interchange
between humans and machines. Form is thus a *relation* between humans and
technology as well as a thing in itself, the medium where infrastructure and
user meet. There can be no technics without form, yet it is separate from those
technics, participating in a paradigmatic chain of relations with previous forms,
their aesthetic histories, and the epistemic worlds that come with them.

Form leads us to the question of political aesthetics—the way that aesthetics, broadly conceived, establishes a political force enabling and contesting
various kinds of authority that circulate in the world. Political rationalities are
fashioned, made palpable, and disseminated through concrete semiotic and
aesthetic vehicles oriented to addressees. The literary theorist Sianne Ngai, for
instance, argues that we exist "in a culture that hails us as aesthetic subjects
nearly every minute of the day" (2012: 23). This aesthetic address is as much a
part of an electricity switchbox, the tangle of cables strung across a street, or
the sound of a generator, as it is an attribute of literature or art. It is certainly
the case that infrastructures are material assemblages caught up in political
formations whose power in society derives from their technical functions. But
they also operate aesthetically, and their aesthetic address constitutes a form
of political action that is linked to, but differs from, their material operations.
And political aesthetics is one way that we can understand the promise of
infrastructures.

Considering the promise of infrastructure allows us to explore the ways in
which infrastructures compress within them different operations and allows

us to focus on nontechnical as well as technical dimensions of infrastructure. Elsewhere (Larkin 2013), I discussed this through the idea of the poetics of infrastructure, drawing on Roman Jakobson's famous parsing of the multiple functions embedded in speech acts. Infrastructures share this compound nature, the potential to operate in different ways and on multiple levels. At times it is their material operations that dominate, the ability to provide electric power, dispose of waste, or create a system for the movement of goods by containers. At other times, as Achille Mbembe and Janet Roitman (1995) have argued, the technical function of infrastructural projects (whether they operate or not) is subordinate to their role in creating a means to transfer public money into private hands. At still other points, governments, leaders, and parties fund infrastructural projects for their sign value, evidence of the ability of parties of the former left to modernize by entering into public-private partnerships, or of municipal authorities to show their commitment to a green, environmental future, or of states to develop society. In the first of these examples, it is the material nature of the infrastructure and its technical function that is paramount. In the second, materiality is a screen for the financial agreements that lie behind it and that transduce technical things into economic things (Mbembe 2001). The last example emphasizes the address of infrastructure. When infrastructure operates in each of these modes it draws together different sets of actors and generates distinctive sorts of political effects.

In this chapter I explore the relation between infrastructure and political aesthetics. I focus on the ways that infrastructures address people as well as move things, how they are composed of form as well as materials. Infrastructures participate in what Jacques Rancière (2006; 2009) refers to as *poiesis*, the act of bringing something into being in the world by creating a way of doing and making, and *aisthesis*, how it is those things produce modes of felt experience. These qualities define infrastructures just as much as art objects, for infrastructures are always fantastic as well as technical objects. They are made up of desire as much as concrete or steel and to separate off these dimensions is to miss out on the powerful ways they are consequential for our world.

Infrastructures, as Stephen Kern (1983) has argued, contribute to our sense of being in time, feeling cut off from the flow of history, attached to the past, isolated in the present, or rushing toward a future.[1] They address the people who use them, stimulating emotions of hope and pessimism, nostalgia and desire, frustration and anger that constitute promise (and its failure) as an emotive and political force (see Gupta; Harvey; and Schenkel, this volume). They express forms of rule and help constitute subjects in relation to that rule, drawing on those measures of hope and pessimism to gain force. Aesthetics are also part of the ambient life that infrastructures give rise to—the tactile ways in which we hear, smell, feel as we move through the world. Political aesthetics refers to both these representational and experiential qualities. Instead of a split between the material and the discursive, or the nonhuman and human, political aesthetics sutures the material and the figural, showing how both are engaged in a constant reciprocal exchange. They make the distribution of rule visible as an aesthetic act. This is why infrastructures are often objects around which political debates coalesce. They are reflexive points where the present state and future possibilities of government and society are held up for public assessment. The promise of infrastructure refers to this political compact, and political aesthetics makes visible the governmental promise of infrastructure as a reflexive, politically charged thing.

Materiality

The rising interest in infrastructure in the social sciences and humanities is part of the more general turn toward materiality (Latour 1993; Bennett 2010; Coole and Frost 2010). Materiality is often taken to be the 'ground' of an object, its most basic, originary condition before the object is caught up into higher levels of discursive meaning. Adrian Mackenzie (2002), for instance, splits the analysis of technology into "layers." One layer is a higher order of meaning in which technology is treated as a historically situated discursive entity representing ideas such as progress or civilization. For Mackenzie, however, there is a second, more fundamental layer that "strongly resists reduction to discourses" (2002: 5).

At this level, technology, Mackenzie argues, is the precondition of thinking, representing, and making sense, not an epiphenomenon of it. There is a causal relation in which the technical is autonomous from and anterior to the discursive which it conditions. Thomas Lemke summarizes this position: "The material turn criticizes the idea of the natural world and technical artifacts as a mere resource or raw material for technological progress, economic production or social construction" (2015: 3). As Jane Bennett puts it, vibrant matter has a life force of its own and "is *not* the raw material for the creative activity of humans" (2010: xiii, emphasis in original). Infrastructures, in this sense, may be introduced as part of a socialist five-year plan, evidence for the superiority of private enterprise over government intervention, or revelatory of the power of Pentecostal churches to remake the temporal (and spiritual) world. But politics, economy, and religion, in this line of argument, represent socially organized discursive realities separate from the primary material level. Conditioned by technics, they cannot themselves condition the technical.

One consequence of the new materialism is to counterpose the material to form, or at least to certain definitions of form. If the problem of Aristotelian hylomorphism was that Aristotle saw form as something imprinted upon matter, reducing matter to a passive receptacle without any agency, materialism has maintained this split while reversing its hierarchy, placing the material as primary.[2] This is why the material turn prefers "unformed" synonyms—matter, material, objects, things—which describe substances in their amorphous, "unformed," elemental state. Infrastructures have an elective affinity with this conception as they are so frequently seen to be a primary technology upon which form is constructed. The infrastructure of a house, for instance, is its wires and pipes, sheet rock and steel, that delimit and make possible the "form" that is laid on top. There is a linear relation here. Infrastructure is primary; form, secondary.

My problem with this split is that it makes it difficult to develop a conception of political aesthetics and form's role in those aesthetics. When I use the term "form," I am drawing on the literary theoretical sense of the imposition of conventional meaning through the formal arrangement of signs. It is about a set of properties a thing possesses—rhyme, rhythm, stress, and meter in poetry; chiaroscuro lighting and oblique angles in film noir; minimalist aesthetics and lack of iconic representation in abstract art and so on. And it is about the sensory effect of those properties on the readers and viewers who engage them.

While I have great sympathy for the emphasis on the essential technicity of the human body and human collectives, I do not see the need to split the technical and the symbolic, insisting on two distinct realms arranged in hierarchical and causal relation rather than as mutually structuring. It also risks fundamentally misrecognizing the range of ways in which infrastructures address, order, and constitute political relations splitting the study of technics from aesthetics and desire rather than seeing these as mutually constitutive. If objects are thought to possess a vital force operating at a level prior to or below consciousness they cannot be theorized in terms of desire, intention, ideology, need, emotion, fantasy, or form—as this would turn infrastructures into what Bennett dismisses as "thoroughly instrumentalized matter" (2010: ix). The promise of infrastructure, however, refers to a political rationality, made up of expectation, desire, temporal deferral, sacrifice, and frustration that takes us into the realm of discursive meaning.

My aim in this chapter is to explore the political aesthetics of infrastructure not through close ethnographic analysis but through a more general theoretical account, moving through a range of examples that draw out the implications of this idea. It is intended to contribute to our understanding of infrastructure by arguing that materiality is simply one of multiple qualities that make up infrastructures, to remind us that technologies are always metaphors as well as technical objects. They bundle together series of things that can be analytically separated but in practice are often wrapped up together and hard to disentangle. The laying of railroads all over Indonesia, for instance, as Rudolf Mràzek (2002) argues, also meant laying down the technical language that went along with those railroads. A railroad was something to be spoken as well as ridden, and railroads came to constitute Indonesian life through language as well as through rolling stock. I will first discuss the introduction of the radio as a new media infrastructure within colonial Nigeria as an example of the reciprocal exchange between machines and ideas and the difficulty of actually separating these into mutually discrete layers. Thereafter I will engage more directly the political aesthetics of infrastructure as a key aspect of infrastructural life.

Radio

In 1939, a letter was sent from the secretary of the Northern Provinces (SNP) of Nigeria to H. O. Lindsell, the Resident of Kano Province.[3] In it, the SNP states he has been directed by the chief commissioner of Nigeria on behalf of the postmaster-general to recommend that radio distribution services be established in Kano city and that to reach African listeners they should install public loudspeakers.[4] He then asked Lindsell if this could be achieved practically and whether the Resident would support such a scheme. The answer was no. One minuter to the SNP's letter pointed out that few Africans would be likely to subscribe, that such a scheme involved large startup costs the province could not afford, and that even if it could with the onset of war there was not the technical staff available to install and support it. The minuter argued it would frankly be cheaper to buy (tunable) radio sets for the few institutions where the intelligentsia gathered rather than build an entire distribution service, but even that could happen only "if they can be bought—at the moment they cannot."[5]

Consequently, the Resident wrote back to the SNP, dismissing the project. Two years later Kaduna (capital of the northern region and the seat of government) pressed Kano once more on the subject. Again the response was negative, and to bolster his opposition the Resident turned to the age-old trick of indirect rule, saying that "neither the Emir . . . nor the Sabon Gari Representative board . . . are in favour of the erection of Loud Speakers."[6]

There the matter lay until 1943 when J. H. Carrow replaced Lindsell as Resident of Kano and almost immediately began to reverse Lindsell's decision.

When Kaduna once again raised the idea of radio broadcasts, Carrow wrote, "I assured Mr Stephens of my strongest support in every way. . . . I do *NOT* consider the opposition to public loudspeakers at key points to be found in previous comments in this file [i.e., by the previous Resident and officials] as in any way final. In my opinion such public loudspeakers will be essential if we are to get at the mass of the population who are illiterate. . . . Radio diffusion is to be installed in Kano immediately."[7] Carrow was opposed to older emirs and colonialists who made indirect rule into a shibboleth that precluded any change and promoted "ISOLATIONISM AND SHELTER FROM THE WORLD" (as he wrote in a later letter, shouting in capitals to indicate his depth of feeling on the subject).[8] Carrow ordered that his thoughts on the issue be distributed to all administrative officers "so that they can read my minute for guidance" and realize "the value [of radio] as a means of disseminating news to a public which cannot read."[9]

I want to pause and think about what is going on here both practically and conceptually as a way of introducing how we might think about issues of infrastructure and materiality. First, I would like to consider the structure of a colonial bureaucracy. When the SNP, acting on behalf of the chief commissioner of the Northern Region, at the direction of the postmaster-general, writes to ask the Resident of Kano a question, this is not just a recounting of a circulatory chain of communication but the recapitulation of a series of authorities designed to induce pressure. Its meta-comment is that many important men at the federal and provincial levels have already come to the decision that radio is an imperative for Nigeria, and while they know they have to ask the Resident his permission, this is not a neutral or disinterested question but a heavily weighted one designed to couch a directive as a question. The fact that they ask again a few years later, and then again a few years after that, indicates the matter is not settled and contributes to the pressure that was building over this issue. It therefore raises the structuring issue of why radio was deemed to be so important?

Second, Resident Lindsell's rejection of radio was driven by the practicalities of colonial rule in a developing society that presented a host of urgent needs, all during a time of war. Radio installation was only one among many infrastructural projects in a developing society. Lindsell faced demands to build more roads, promote agriculture, erect hospitals, improve the police force, promote industry, and broaden education. Given that his funds could cover barely a fraction of these needs, the demand for a commitment to radio necessarily meant a weakening of his obligations to other areas. Even if the funds were available to support radio, many technical personnel were away at war and those remaining would have to be seconded from different departments. Raw materials were scarce, heavily rationed, reserved mainly for the war effort, and demanded by different departments. In the mind of Lindsell and the officers below him, radio was a luxury, far down in the list of priorities for Kano Province, and they felt that technical staff, raw materials, and capital should be directed elsewhere.

To build a networked infrastructure such as radio distribution involved prioritizing radio over a host of competing claims. Lindsell's refusal and the support of his junior officers in this refusal (as well as resistance from different Nigerian communities) give a sense of the intensity of this opposition and the entrenched situation that Carrow faced. To counter these claims, Carrow had to argue why radio should have priority over roads, increased electrification, water supply, or road building. This meant defining an argument that made radio installation a priority; advancing that logic in meetings, minutes, letters, and circulars; establishing its authority; and making it hold sway in a competitive environment where counterarguments were present. Carrow ordered his minute in support of radio be distributed to all administrative personnel because he realized that the junior officers underneath him likely did not support his position and that therefore he needed to instruct them in the new priorities they were to follow.

Carrow's actions were driven by a logic of governmental rationality, which provided the exterior conditions of existence for radio. Without it, no copper wire was imported, no poles erected, no personnel trained. The materiality of radio as a technological ensemble—its microphones and speakers, amplifiers, electric wires, and telephone poles—only came into being because of nonmaterial arguments that governed their existence, regulating (but not fully controlling) how that ensemble operated in the world. Once those systems were built, their operations could not be fully controlled by the political rationalities that went into their funding, but their evolution depended upon a constant reciprocal interaction between the technics of a system and the external conditions (forms of political rule, modes of capital, religious structures) from which technological systems emerge.

Carrow's support for radio can be encapsulated by the phrase "the promise of infrastructure." A promise can refer to a vow, or a commitment, but its other meaning refers to the coming to be of a future state of affairs, the idea we have that someone or something holds promise. Its referent is not to the here and now of things but to an uncertain future that infrastructure is to bring about and institutes a temporal deferral that refuses to deliver something in the present.[10] It involves both expectation and desire, frustration and absence. It calls into being a future world that is at once planned for, administered, and organized, but also made up of a longing that is not always ordered by rationality. Infrastructures, in this sense, are promising technological ensembles. The very word "promise" implies that a technological system is the aftereffect of expectation; it cannot be theorized or understood outside of the political orders that predate it and bring it into existence.

To read the files concerning radio's introduction into Nigeria is to be fully saturated by the promise of infrastructure. As I argue in my book *Signal and Noise* (Larkin 2008), radio in colonial Nigeria represented a particular mode of political rationality whose aim was to fundamentally reshape the everyday practices and sensibilities of colonial subjects.

Carrow saw the medium as one of the "modern aids to progress," which, by circulating information and exposing what he saw as a backward people to alternate ways of life would have the cognitive effect of loosening them from traditional lifeworlds (what Carrow referred to as their "parochialism") so that they could begin to imagine other ways of living. Radio connected Nigerians to broader worlds. Its promise was that it would promote a "progressive interest in affairs happening outside the community" and make people "more knowledgeable and curious of events further afield."[11] For British colonialists like Carrow, it was a machine that operated upon people's cognition, forging new social subjects, and they hoped to mold those subjects according to the priorities of colonial rule.

In his conception of media, Carrow stands in an august tradition of media theory that makes for odd bedfellows such as the Marxists Walter Benjamin and Siegfried Kracauer, the modernization theorist Daniel Lerner, and modern-day avatars as diverse as Arjun Appadurai and Friedrich Kittler. Each believes that the technicity of a medium has cognitive effects, that it structures perceptual abilities and transforms consciousness—or, as Friedrich Kittler has it, media "culturalize the natives" of any society (1987: 159). Carrow advocated radio while those around him (the Residents of Sokoto and Bornu, for instance) rejected its necessity. "There is always the tendency for Sokoto, Gwandu and Argungu to be parochial and wholly disinterested in matters outside their own area," Carrow wrote to the SNP in 1944. "If this cannot be countered then the Sultan and the Emir of Gwandu cannot possibly . . . spur forward the peoples inhabiting their Emirate."[12] The expectation was that radio could bring about forward-thinking people, unleash the forces of progress, and remake subjectivities—and with that open up the circulatory forces of liberal capitalism. Promise is not the posterior encoding of a material assemblage into discursive meaning, nor is it the anterior condition of possibility from which radio as a technical system emerges. It is both. In Simondonian terms (1958), to look at promise is not to examine one state or the other but to examine the sets of relations between states, the ongoing dynamic exchange between different elements (an internal technical tendency and an external milieu of political rule). Radio is part of the sets of exchanges in which the technical system, the human, and forms of political rationality mutually constitute one another.

To insist on the importance of the political address of the medium, and the way it is caught up in a complex ecology of colonial rule and cultural and social formations, does not deny the autonomy of the material. In an earlier work (Larkin 2008), I pursued this question by raising two points. First infrastructures are not just technical but also conceptual objects governed by exterior political conditions that form the conditions of possibility for their emergence.[13] In colonial Nigeria before a radio program was listened to, before a road was driven upon or a tap opened, there was a confrontation between subject and technology whose stakes were contested—by colonial authorities, nationalist leaders, Islamic preachers—and which were bound up with forms of desire, anxiety, promise, and fear. At the same time, I argued that infrastructures are not just conceptual but technical objects whose material operations engender wholly new conditions of existence outside of the political imagination of those contesting them. The technical operation of radio was dramatically affected by the physical life it led in Nigeria where it interacted with white ants, humidity, and harmattan dust that often caused radio components to fail. These material operations were not under the control of human design but part of the unexpected contingencies of all that can happen to machines in reality. But that reality was also shaped by the predicates of Islamic law, the ambitions of colonial rule, shortages of raw material and personnel, and the theories of media that lay in the minds of modernizing colonial officers.

As a concept, promise is tied to the political aesthetics of infrastructural systems. These do not have just technical requirements—circulating radio waves, vehicles, people from one place to another—but transmit ideas at the same time. Those ideas address people, create subject positions—deeply attractive for some, repulsive for others—through which they operate to fashion sensibilities. Taking all of these into account allows us to expand our concept of infrastructure, to draw on the insights gained from the material turn but without rejecting the fact that infrastructures are also figures.[14] It is precisely because infrastructures are invested with promise and because that promise is reflexively foregrounded that—when they work or when they fail—they bring into visibility the operation of governmental rationality and offer that rationality up for political debate. To understand how this takes place, how the material and the figural are brought together, it is useful to draw upon the technical device upon which political aesthetics rests. For the rest of this chapter, I am going to explore this concept in relation to infrastructure, particularly through the idea of form, as a means to aid how it is we "think infrastructurally" (Chu 2014: 353).

Political Aesthetics and Form

As a conceptual term, form is diverse, with deep philosophical, literary, and epistemological references and mutually exclusive meanings. It is both a noun in that it refers to concrete things and a verb in that it denotes the ordering and forming of things (referring back to its older meaning of the mold upon which wood or metal is shaped). As Caroline Levine (2015) argues, "Form can mean immaterial ideas, as in Plato, or material shape, as in Aristotle. It can indicate essence, but it can also mean superficial trappings . . . mere forms" (2). Form can be abstract or highly particular, "cast as historical, emerging out of particular cultural and historical circumstances or . . . ahistorical, transcending the specificities of history" (2). For some, form is the human process of arranging patterns while for others it is an emergent property of the natural world that operates outside of human intervention (Kohn 2013).

At its broadest level, then, form is a matter of ordering. It is about the structuring and patterning of experience, imposing order on the world, and, at this level, refers to a wide range of artistic and social phenomena. Form forms things. It operates upon people and makes them into particular sorts of subjects. It does so through various operations. Form is representational in that it addresses people, distributing particular sorts of political rationalities whereby, for instance, the state can seek to impose its sense of the world and citizens accept or contest that ordering. These are historically constituted, complex projects of political sovereignty that, in order to be effective, have to be communicated as effects of rule. Forms also create phenomenal experience. They are part of the interface bringing technology and user into productive engagement. Brandon Hookway (2014) argues that an interface, which seems to be a technological object, is more properly a way of engaging with technology. It comprises the bottleneck through which all human relations with technology must pass. Because of this, form is a concrete thing that must be used: picked up, held, walked upon, sat on, turned, or pulled. It provides physical experiences and creates a sensory, tactile environment that translates political rationalities into ambient experience.[15]

For some anthropologists, form is dominated by its association with art and literature and seemingly irrelevant for a mass of other phenomena. Posters, cathedrals, sonnets, and sculpture have form, according to this thinking, while traffic lights, pipes, prepaid phone cards, roads, and cables do not. For some, a turn to form is a turn toward the surfaces of objects, from which ideological, political, or cultural processes can be "read" and which stand in contradistinction to the material reality that stands underneath and conditions those forces. This binary is often blurred in practice and I am precisely interested in the formal qualities of everyday infrastructures and their role in producing a political aesthetics. This is not a split between the material and the representational, as form is as concrete a thing as pipes and sewers. This argument goes back to the Russian formalist rejection of symbolic analysis and "speculative aesthetics" in favor of tangible, observable technical devices. As Boris Eichenbaum characterized it in a 1926 review essay, form is not "the outer covering but something concrete and dynamic, substantive in itself" (2004: 9). Literature, Eichenbaum flatly stated, is "a specific series of facts" (12).[16] The second major argument that the formalists developed is that these concrete devices are arranged in order to create sensible effects on people, what Victor Shklovsky termed the "means of creating the strongest possible expression" (1965: 8). This is a mode of poiesis, the use of form to bring something into presence, and aisthesis, the felt experience of that thing. For Shklovsky, form induced a sensorial effect. Its function was to "remove objects from the automaticity of perception," bringing them to visibility so that their sensuous qualities were reflexively experienced rather than fading into the background (13).

What we can take from these ideas is that form has a concrete thingness that is in complex reciprocal interaction with the material properties from which infrastructures are made. Second, these forms impose sensory conditions of experience. While wholly uninterested in the political conditions from which literature derives, Russian formalists were centrally concerned with *effect*, the sensorial experiences that forms provoke. Precisely because infrastructures are caught up in relations of the state to its citizens, form translates modes of rule into concrete visible structures, making them affectively real and emotionally available.[17] Form induces cognitive and affective dispositions—an argument that Ngai sees as constitutive in forming the political experiences of subjects. It is similar to what Raymond Williams referred to as structures of feeling—the particular quality of social experience that is produced by dynamics of historical change, or, as Williams has it, "meanings and values as they are actively lived and felt" (1977: 132).

One can argue that the question of the political aesthetics of infrastructure is intimately related to the vexed issue of the visibility or invisibility of infrastructures. It has often been argued—almost typically argued—that infrastructures comprise the invisible, taken-for-granted substrate that allows our world to operate (Star 1999; Graham and Marvin 2001). The critical examination of infrastructure, in this light, is to perform an act of "infrastructural inversion" (Bowker and Star 2000; Peters 2015) by bringing what is background into the foreground. In many instances this is clearly correct and a powerful intervention. But infrastructures are not normatively invisible and then brought into visibility by some sort of exceptional act. Visibility and invisibility are not ontological properties of infrastructures; instead, visibility or invisibility are made to happen as part of technical, political, and representational processes. This is why the distinction between spectacular infrastructures and mundane ones should not be figured as an opposition but as representing different styles of visibility. When technical systems are removed from public discussion and become the preserve of experts, for instance, this can be seen as a *practice* of occulting, just as the demand for greater transparency around infrastructural projects is part of the work of visibilization. In her study of the introduction of water meters to South African townships, for instance, Antina von Schnitzler (2016) argues that precisely because apartheid was a project organized through infrastructural segregation, antiapartheid activists tended to take infrastructures as the foci around which protest could be organized. To move away from this protest, von Schnitzler argues, postapartheid governments introduced new technological devices such as prepaid water meters as an explicit means to turn what had become a political relation into a neutral, technical one (something she terms the logic of administration). What von Schnitzler shows, here, are active attempts to invisibilize infrastructures just as the mass protests that occurred around them can be seen as a refusal of that occulting. Visibility or its opposite is not an inherent quality of infrastructures but practices whereby politics is struggled over.

Infrastructures represent and are represented in their built forms; the protests that congeal around them; the sets of numbers, graphs, and tables by which they are administered; the budgets that undergird them. These depend on various material and formal devices, each of which invokes specific modes of address, draws together specific sets of actors, involves differing uses of secrecy and transparency, and constitutes the political in distinct ways. Conceived of in this way, the concept of political aesthetics I lay out shares a great deal with Rancière's definition of the politics of aesthetics. While Rancière is primarily concerned with works of art and largely uninterested in popular forms, let alone infrastructural systems, his understanding of the role of aesthetics in the constitution of political subjects offers an important way into thinking about the significance of infrastructures.

Rancière

Rancière does not see aesthetics as a domain that is opposed to politics but as the means through which the political is constituted and operates. Politics, for Rancière, is not about the struggle for power but is "the configuration of a specific space, the framing of a particular sphere of experience, of objects posited as common and as pertaining to a common decision, of subjects recognized as capable of designating these objects and putting forward arguments about them" (2009: 24). Politics takes place when those who occupy fixed positions outside a certain order decide to intervene within that order. It is the apportioning that determines who can participate in a system: "This distribution and redistribution of places and identities, this reapportioning of spaces and times, of the visible and invisible, and of noise and speech is what I call the distribution of the sensible" (2009: 24–25). His aim is to expose the categorizations that assign objects to a specific place and define who has the ability to speak about them. What links aesthetics and politics, for him, is that both participate in constituting these categorizations, and both share an ability to bring into being worlds and to interrupt the distribution of those worlds. Aesthetics, he argues, is the system of "*a priori* forms determining what presents itself to sense experience. It is a delimitation of spaces and times, of the visible and invisible, of speech and noise, that simultaneously determines the place and stakes of politics as a form of experience" (2006: 13).

If art shares with politics the power to constitute new collective worlds, Rancière argues that this occurs through processes of poiesis and aisthesis. These are the domains that are, for me, of the most interest for the study of infrastructures. Poiesis, in this context, refers to the process of doing and making, the techniques whereby a broad range of things are brought into sensible existence. Aisthesis is the sensory apprehension of those things and the world they create. "Aesthetic acts," Rancière argues, are "configurations of experience that create new modes of sense perception and induce novel forms of political subjectivity" (2006: 10). For Rancière, the politics of aesthetics is the role that art plays in making visible the distribution of sensible order and offering a critical alternative to it. He is at great pains to distinguish this from the staging of mass spectacle in fascism that Benjamin saw as crucial to the aesthetics of politics. Instead, Rancière is interested in the aestheticization of common experience, the distribution of order between subjects and the world, and the ways those subjects dissent from that distribution. He sees art's political role as opening up spaces for transformation and disruption, creating new collective spaces from which the consensual order can be challenged by posing alternatives.

My interest, by contrast, is in the aesthetic operation of everyday infrastructures, how it is that anonymous infrastructural phenomena—switches, pipes, cables, roads, sewers, bridges, railways, servers—operate on the level of form as well as technics. Their political significance lies in these formal operations as well as in their functions. Rancière does not examine how common objects have aesthetic force, instead adopting a relatively familiar depiction of the artist as critical outsider, proffering alternatives to the social order. One of the great ironies of Rancière's work is that while he makes a theoretical argument for breaking down the distribution that separates art from ordinary life, the vast majority of the references he draws upon to do so come from high literature and art. For instance, in *Aesthetics and Its Discontents* (2009) he cites a passage from Stendhal's *The Life of Henry Brulard* (1958) in which Stendhal recalls the sounds of his youth—the church bells in the evening, the sound of servant girls using a water pump, the playing of a flute by the apprentice—that brought an awakening appreciation of music into his life. This passage is significant for Rancière because it conflates a world of artistic achievement (the flute) with the machinic sounds of the pump, placing both onto the same aesthetic plane. "Far from demonstrating the independence of aesthetic attitudes with respect to artworks," Rancière argues, "Stendhal testifies to an aesthetic regime in which the distinction between those things that belong to art and those that belong to everyday life is blurred" (2009: 5). However, this richly suggestive sentence opens a perspective that Rancière never follows except at the most general level. He does not extend his analysis of aesthetics into the infrastructural realm (the reference, after all, is to Stendhal, not to water pumps). I wish to bend Rancière in order to see infrastructures as sharing with works of art the similar role of producing sensory experience and through that experience constituting political life. Thought of in this way, infrastructures are formal expressions of experience, vehicles whereby that experience is made palpably real to people, and it is because of this that they are so often the places where public controversy about the shape, nature, and direction of historical change becomes publicly available and debated.

We can compare Rancière, for instance, to the far richer discussion of water pumps in Mandana Limbert's *In the Time of Oil* (2010). Limbert examines infrastructural development that came to Oman after the advent of oil money. One aspect of this was the revolutionary state's provision of piped water, which entered directly into the home and replaced the use of communal wells and of timed irrigation. For Limbert, the well stands for past time; it is a ruin in Benjamin's (1999) sense, distilling a social order and mode of production that can no longer be supported and which has been replaced by motor pumps. Oman, she tells us, "falls squarely between naturalized assumptions that development, in the form of piped water, mechanical pumps, and sprinklers, is necessary for the fulfillment of necessary statehood and notions that older forms of water distribution are emblems of Oman's past values and knowledge" (2010: 117). Water supply, as she describes it, becomes a meta-reflexive sign of the loss of a past lifeway and the anxious possibility of the future, and it is so emotively powerful that one farmer asks her to measure his well and take pictures of it so it can be re-created in a museum. She describes another farmer who tape-recorded the sound of his well so that, unlike Stendhal, he can listen to the sounds of his youth in a world where those sounds have been replaced by the hum of electric motors.

Infrastructures are tied to the political conditions that govern their existence and the emotional entailments generated by those conditions. As public goods, they represent a relation between a state and its citizenry, and they are embedded in what Laura Bear (2015) terms the *res publica,* an idea of care for the world that is as ethical as it is political. States are expected to provide a certain level of care for their citizens, what Achille Mbembe (2001) refers to as the production of public happiness. When the state does not seem to be living up to its agreements, when infrastructures fail or are not completed, the intensity of response and anger is driven by the affective politics that result from those ethical obligations. Akhil Gupta (this volume) describes this as "the biopolitical project of creating citizens who share the goal of inhabiting a modern future." Infrastructures never just supply electricity, water, or gas. They implicate the very definition of the community, its possible futures, and its relation to the state. When Rancière argues that the distribution of the sensible "defines what is visible or not in a public space" (2006: 12), part of what it is making visible and holding up for public discussion is the nature of these ethical obligations around which infrastructure is turned into politics and politics into a structure of feeling. And while Rancière is right to see aesthetics as a specific form of sensory apprehension, he does not explore how this is used by the state to distribute forms of political rationality and to create ambient environments in which that rationality is experienced. It is because they are so strongly associated with forms of political order and their dissolution (or refiguring) that aesthetic objects have the capacity to become metapragmatic objects, signs of themselves, deployed in particular circulatory regimes to establish sets of effects that dramatize political conditions and make them subject to public debate.

Figure 7.1 is a photograph by the Magnum photographer Bruno Barbey of a new highway in Lagos, Nigeria. Barbey's image depicts a road that promises free, uncluttered movement, a way to assuage the desire and fantasy of mobility. As Armand Mattelart (1996) has argued, this coding of circulation derives from the Enlightenment logic that humans should not live in fixed states but that both individuals and societies should be open to change and mutability, and that progress is brought through the free circulation of goods, ideas, and people (the exact episteme mobilized by Carrow to justify funding a radio network). A variant of these ideas came to be formalized as modernization theory but its logic extends far back into Enlightenment liberal thought, perdures through colonial rule, and extends past the era of decolonization. Both colonial and nationalist governments were intensely modernist, fully adopting the logic that it is through infrastructures of circulation that development and modern subjectivity can be achieved. This is why roads became for both colonial and nationalist governments and their peoples the defining object through which development could be pursued.[18] Roads are both the technical means to bring about development and signs of that desire for development. They create both a physical space that people must traverse and a mode of address by which those people are interpellated.

We can see some of the origins of the ideological charge encapsulated in Joyce Cary's classic novel of district administration, *Mister Johnson* (1939). Set in northern Nigeria, *Mister Johnson* tells the story of a clerk, Johnson, brought in at a low level to the bureaucratic system of colonial rule from where he sees his European superiors embark on a series of development projects. The book follows two stories: Johnson's inability to manage his money, and the embezzling for which he will ultimately be caught, the aspect for which the book is most famous (and most criticized for repeating the cliché of corrupt Africans). The other story, however, is far stranger and follows Rudbeck, Johnson's superior, an assistant district officer in the northern Nigerian town of Fada, who is obsessed with road building. Rudbeck, Cary tells us, is a man with a "passion for roads" who has "caught the belief that to build a road, any road, is the noblest work a man can do" ([1939] 1989: 46).

Cary depicts Rudbeck alternately as faithfully trying to discharge his responsibility as a colonial officer to bring development to the region, and as somewhat unhinged and fetishistic in this ambition. When he writes that Rudbeck wishes to "build a road, any road," it suggests an unrestrained desire for civilization without regard to practical purpose. Moreover, Rudbeck is not alone in

FIGURE 7.1 An empty motorway in Lagos, 1979. © Bruno Barbey/Magnum Photos.

this belief because he "caught it" as if it were a contagion, from his own superior, Sturdee, and now seeks to reproduce these ambitions to the extent that he has slightly lost touch with reason.[19] It mimics Carrow's faith in the modernizing effect of the radio that takes place without regard to the content it relays. The technical specificity of the road (which areas it connects, how it moves people) seems less relevant than its promise as a moral and civilizational tool.

Cary himself was a colonial officer, based in the north of Nigeria, who caught his own enthusiasm for road building from his superiors (the character Rudbeck is a loosely fictionalized version of Cary). Cary saw in roads, and communication more generally, the civilizational promise of liberalism. "The first need in Africa has always been communications," Cary wrote, in a collection of political essays: "Trade, order, peace, the intercourse which comes from trade and which is the very beginning of civilization and the education of people all start from the free and safe harbor, the open river and the cleared road" (cited in Moody 1967: 146). In a letter to his wife while road building in Nigeria as a district officer, Cary gave a more emotional sense of how this civilizational promise was felt: "I am starting a new grand trunk road. . . . I cannot explain the pleasure of seeing a road which one has planned and surveyed in actual being" (cited in Moody 1967: 146). Cary describes how the road vivifies a particular form of political order but the way he describes its power often seems to stray from the rational to the fantastic. In *Mister Johnson* this extends so far that the road becomes a literal fetish as Cary depicts it taking on animate form and talking to Rudbeck or, depending on how you read it, shows Rudbeck losing touch with reality and hearing voices coming from the road. "I'm smashing up the old Fada," the road tells Rudbeck. "I shall change everything and everybody in it. I am abolishing the old ways, the old ideas, the old law; I am bringing wealth and opportunity for good as well as vice, new powers to men and therefore new conflicts. I am the revolution. . . . I am your idea. You made me" ([1939] 1989: 168, 169). "I am your idea. You made me" encapsulates not just the ways that the aesthetic address of the road comes to overwhelm rational technics, but it neatly vivifies the reciprocal exchange between idea and material thing.

The belief in the power of infrastructural development was, if anything, an even greater part of nationalist struggle than of colonial rule. This was especially the case in Nigeria after the oil boom of the 1970s, which ushered in what Michael Watts (1992) has referred to as a frenzy of infrastructural building. Federal and local states sought to invest in infrastructural ventures as a means of developing society and as a political technique of displaying state power through what Fernando Coronil described as the "theater of modernization . . .

dazzling modernization projects that engendered collective fantasies of progress" (1997: 239).[20] Barbey's book on Nigeria mostly concentrates on cultural
and religious rituals, traditional architectures and dress, but he juxtaposes this
"tradition" with images depicting oil platforms and roads—icons of 1970s
modernization. The road appeared to Barbey as an object to be photographed
precisely because it stored and represented the promise of infrastructure making that promise—as a mode of political rationality—emotionally real.

It is this political rationality with its admixture of technical rationality
and fantastic excess that is engineered into the Lagosian highway that Barbey
depicts, and it forms part of its aesthetic effect. The formal qualities of the road—
its blackness and hardness, the swooping clover leafs and graceful curves—
generate an address that calls forth specific subjective capacities and emotional
experiences and provides a way of relating those experiences to broader social
arrangements.

Barbey's road can be contrasted to another, perhaps more familiar, depiction of contemporary Lagos (figure 7.2) and the crowded congestion that has
come to mark urban Nigerian life. The road conditions in this second image
draw their political charge from the betrayal of the promise that was offered
in the first. The collapse that followed the oil boom revealed it to have been a
period of excessiveness and irrationality and infrastructures became icons of
that irrationality—massive state projects that hemorrhaged state funds but
which were often incomplete, poorly constructed, and subject to constant
breakdown. Barbey's road stands as an infrastructural promise of modern
development because it is tied to structural shifts in Nigerian economy and
society. From the perspective of the present, these earlier ambitions are revealed to be a desire for a futurity that was as fragile as it was intense. The
betrayal of that promise becomes the grounds for debating what that political
promise was and what caused its failure. Every traffic jam, every pothole, every
incomplete road becomes a means by which the state is brought to public attention. It sets in motion an everyday hermeneutics about why infrastructures
fail—corruption, incompetence, ethnic favoritism, or any of the other litany
of reasons commonly advanced to explain a contradiction that people feel
keenly. Political aesthetics captures these dimensions. It refers to the representational work of form in distributing political rationality but also to how that
rationality is sensed through the ambient environment and the felt experiences that roads generate. Those who use the road are subject to the ordering
of those ideas through the physical experience of engaging its space.

Barbey's photograph depicts the highway that connects mainland Lagos to
the islands that host the markets and elite business and residential areas of the

FIGURE 7.2 Congestion in Lagos. Photo by the author.

cities. Congestion on highways and roads is so bad that Lagosians time their entry and exit from work to avoid the notorious "go-slows" that can snarl traffic for hours. The regularity of congestion has turned roads—at certain times of the day—into spaces of sedentarization rather than movement (Amiel Bize [2017], writing about Nairobi, refers to this as "jam-time").[21] Go-slows are both things in the world and events that create the platforms for other actions to happen. They give rise to new modes of planning and new forms of behavior. Ordinary workers wake early to travel, then stay late and eat near work to avoid the postwork rush hour. One wife I knew drove from her home on the Lagos mainland against the flow of traffic to eat dinner with her husband near his work so that they could spend time together while the traffic dissipated rather than having to eat apart. These are ephemeral ways in which the realities of infrastructural life impress themselves upon quotidian existence. They indicate both the possibility of circulation, its foreclosure, and the ability to overcome that foreclosure by finding a way around it or waiting it out.

Go-slows have become as much a ubiquitous symbol of contemporary Lagos as the Eiffel Tower is for Paris or the Statue of Liberty is for New York, commented on by all visitors, and a daily topic of concern for Lagosians. They have created a contested fulcrum around which the state of urban life and of the condition of society is debated, everyday allegories about the state of things. Though quite what the jams mean is contested. On the one hand are those who see the jams as the iconic example of the disintegration of the state, part of an apocalyptic depiction of Lagos and Nigerian urbanism.[22] On the other hand are those who view transport infrastructure and its failures as constitutive of a creative informal economy: "thriving with entrepreneurial activity" (Koolhaas et al. 2000: 674). Lagos has become, perhaps, the most discussed, photographed, reviled, and celebrated city in Africa, a condensed signifier for the state of African urbanism and its uncertain futures.[23]

Barbey's empty road in figure 7.1 or the congested one I photographed in figure 7.2 are not neutral depictions of phenomena in the world but metapragmatic signs about order, time, futurity, chaos, backwardness, and the modern.[24] These ideas are the conditions of possibility for the emergence of the road, just as the physical life that the road leads reshapes what those ideas are and how they gain political force. There is a constant, dynamic interaction between the two out of which both evolved. The fact that Barbey chose a newly constructed road to photograph (one of several in his book) and I chose a congested one is because they present themselves as sensible to us, already overburdened with political freight. They appear as objects "to-be-photographed" precisely because both of us recognized the address of the road and its relation to sensible politics. This is, of course, the ordering mechanism that the distribution of the sensible enacts. It captures me just as I capture it in my camera.

Aesthetic form is one of the technical means by which promise as a political technique is enacted. Form generates a mode of address that induces affective and cognitive dispositions, distributing political rationalities (its representative function). It also constitutes physical environments whereby those rationalities are experienced as lived practice. Itself the outcome of ordering, form orders those subject to it. And all of these various operations take place in an encounter that fuses the material with the human, that does not push beyond the human but reveals how the encoding of social relations is a central part of the material operation of infrastructures.

Conclusion

Naveeda Khan, writing about the development of the Lahore-Islamabad motorway, "the first "American-style motorway ever built in the Indian subcontinent" (2006: 87), sees promise as constitutive to its existence. Khan tells us that for Pakistanis the motorway promised economic development and cultural integration (94), the ability to revolutionize communication (88, 102), "a tightly networked exchange system between the state and the travelers" (102), and, not least, "speedy and safe travel" (100). She also charts how the actuality of the motorway delivered on some of these promises, but abjectly failed on others: "As a communicative technology its [the road's] promise hovered over its actuality. It was saturated by the state's presence even as the state went into partial eclipse with the failure of its circuitry" (105).

Khan highlights here the fact that infrastructures operate at different levels at the same time. She parses a technical object—a road—examining it in terms of form. It is an *American*-style motorway, counterposed to other sorts of motorways potentially available to Pakistanis and starkly differentiated from the older trunk roads built by the British in the nineteenth century. Khan sees the desire for the motorway as part of the distinct preference amongst Pakistani elites for things American. No doubt this preference is dialogically related to a move away from aspiring to things British, which might both be a nationalist sentiment and a recognition of the particular authority of the American hegemon. In any case, "American-style" places the motorway into a relationship of form with other American-style things—from fast food restaurants to strip malls—none of which participate in the materiality of roadness. But the promise of Americanness released by the ambition of the road is part of its structural nature, as immanent to it as concrete pylons.

For some materialists, to argue that infrastructures "represent" forms of capital, or congeal social relations and social labor, is an error because it makes objects into the passive receptacles of human categorization: "white screen[s] onto which society projects its cinema" as Latour phrased it (1993: 53). In this regard, new materialism is opposed to older forms of historical materialism. "Things lie," Henri Lefebvre argued, and as commodities they do so in order to conceal the social labor that goes into their formation. "The unmasking of things to reveal social relationships is one of Marx's great achievements" (1991: 83), he argued decisively. It is precisely this unmasking that science studies and new materialism finds so troubling because it threatens to recenter the human subject as the sole locus of agency. It assumes matter's plasticity or passivity and reinforces the idea that, as Diana Coole and Samantha Frost argue, "matter is inert stuff awaiting cultural imprint" (2010: 26).

As is clear by now, I reject this separation of the material from the discursive and from form. To recalibrate Latour, this is a purification that insists on a split between the human and the material, something he warned against. But while Latour (1993) insists on the material agency of objects, dismissing the idea they were simply shapeless receptacles of social categories, my emphasis is the reverse. Objects are not simply material assemblages wholly autonomous from aesthetic fields and the political rationalities that accompany them. Instead, there is an energetic exchange between the two whereby external environment and technology interact, each shaping the other. As the example of the introduction of radio to Nigeria suggests, analytically insisting upon the medium as a material assemblage operating at a prediscursive, affective level is an impossible analytic act that can be achieved only by purifying radio of the deep epistemologies of colonial rule, cognitive mutability, and theories of media influence that are its conditions of possibility. It remains important to recognize that technology has a material excess that cannot be fully reduced to the sets of ideas that administer it, but that does not mean that those ideas are absent from it at the level of organizing its material presence.

One of the most exciting sides of new materialism is its emphasis on emergence and the becoming of matter rather than its fixed ontology (Barad 2007; Coole and Frost 2010). I see the dynamics of encounter as formed through the constant evolution of objects in relation to discrete environments, environments that are at once physical, political, and social and that take in legal and religious domains as well as the internal, technical logic of machines. This is not a restatement of human exceptionalism or a denial of the autonomy of the material. I recognize that there is an excess to the material that cannot be fully contained by discursive regimes but I also see form and aesthetics as part of the material as well as the discursive world. At times, as Eduardo Kohn has elegantly shown, form can be nonhuman, produced in naturally occurring situations as a "constraint upon possibility" (2013: 157). But the human operations of form are equally important and an integral part of the political aesthetics that are a constitutive part of our world. Form and aesthetics are ecological as well as representational, involved in creating concrete environments as well as addressing subjects. The relationship between the material and the figural—particularly in the case of infrastructures—is reciprocal and entangled rather than causal and successive. Matter and form are present at the self-same time, mutually shaping each other. Technical objects cannot exist without both, and both are essential to an understanding of political aesthetics.

NOTES

My chapter benefited greatly from comments received at a School of Advanced Research seminar. I thank all participants for their thoughtful responses and particularly Nikhil Anand, Hannah Appel, and Akhil Gupta for convening the intellectual ideas we explored. I also benefited greatly from discussions with the Chicago infrastructure reading group and would like to thank Julie Chu, Michael Fisch, Eleana Kim, Jun Mizukawa, and Bettina Stoetzer for their insights. Peter Connor, Meg McLagan, and Jesse Shipley read and commented on earlier drafts and their feedback contributed greatly to the final revisions. Amiel Bize and Rafaella Schor helped greatly with copyedits.

1 This argument is amplified in the introduction to this volume and in the chapters by Appel, Harvey, and Gupta.

2 For a critique of Aristotle see Simondon 1958, Mackenzie 2002.

3 The SNP was the highest official in the Northern Region of Nigeria (NR), one of the three semiautonomous regions that made up the Nigerian state. Nigerian National Archives, Kaduna (NNAK)/Kano Prov/4364/Radio Distribution Services.

4 Radio Diffusion was the wired relay of radio broadcasts to individual subscribers. Operating somewhat similarly to cable television today, it did not depend on a broadcast but instead wired the signal directly to receiving sets that could not be tuned to other stations. This allowed the British to control which stations Nigerians could listen to. See Larkin (2008).

5 Minute, GRJ, 27/9/39; Letter from SNP to Kano Resident, NNAK/Kano Prov/4364/Radio Distribution Services.

6 Kano was a centuries-old city ringed by a mud wall. The onset of colonialism brought with it southern Christian migrants who lived outside of the traditional city in a new area called the Sabon Gari. The Resident is thus emphasizing that all stripes of native opinion were against the introduction of loudspeakers and a radio distribution service. During a certain period of British rule in the north, evoking the dissent of the local population was enough to prevent any administrative effort. NNAK/Kano Prov/4364/Radio Distribution Services.

7 NNAK/Kano Prov/4364/Radio Distribution Services, Note, Resident Carrow 29/6/43.

8 Rhodes House Mss.Afr.s.1489, Papers of J. H. Carrow.

9 NNAK/Kano Prov/4364/Radio Distribution Services, Note, Resident Carrow 29/6/43.

10 See also the introduction and also Harvey, this volume.

11 NNAK/Kano Prov. 4364/s.13, Circular, D. B. Wright for Ag. Civil Secretary, Kaduna to Resident Kano, 28/8/52.

12 NNAK/MIA/765 Radio Diffusion Service, NR.

13 Here I drew on Foucault's argument in *The Archaeology of Knowledge* (1972).

14 Schwenkel, this volume, offers a rich example of this process in her discussion of the smokestack as icon in socialist Vietnam.

15 Siegfried Giedion ([1928] 1995) provides an example of this when he writes about the new experience of space produced by the rise of iron technology, the first building material produced by an industrial process rather than nature. Iron's capacity to bear weight allowed for the use of thin pillars, producing a heretofore "unknown transparency, a suspended relation to other objects . . . [a] sensation of being enveloped by a floating airspace while walking through tall structures" (102). It promoted "freedom of circulation, clear layout, and . . . [permitted] the best utilization of light" (117). Nineteenth-century railway stations, festival halls, or department stores were not just visual expressions of transformations in capital, but, for Giedion, they created a phenomenal experience of moving in space and perceiving light. This experience was both an expression of the age and also a means by which that meaning was physically impressed upon people. Airspace was a physical, ambient experience that for Giedion both emerged out of, and could stand for, structural shifts in society. It was both concrete thing and metaphor.

16 Eichenbaum referred to this as the principle of palpableness. See also Jakobson's discussion of the palpability of the sign in his famous definition of poetics (1985).

17 Penny Harvey and Hannah Knox argue that "even the most unlikely infrastructural projects are able to sustain an ongoing emotional charge" (2015: 28).

18 See Harvey and Knox (2016) for an example of the link between modernization and roads in Peru. Dimitri Dalakoglou (2017) argues elegantly that the Albanian state used roads to surface the country with the ambitions of socialism—but he, Harvey, and Knox are all well aware of the fraught outcomes of these ambitions.

19 As I note in my book *Signal and Noise* (2008), this slippage between rational achievement and fetishistic irrationality over infrastructure development marks the film *Bridge on the River Kwai*.

20 See also Apter (2005), Watts (1992).

21 In an extended discussion of jam-space and jam-time, Bize argues that they are a "rhetorical vehicle through which the vicissitudes of Kenyan society and urban life are discussed so people not only spend time *in* jams they spend time talking *about* them" (2017: 60, emphasis in original). Caroline Melly (2017) explores this beautifully in her discussion of the *embouteillage* (bottleneck) in Dakar. She argues that embouteillage is such a ubiquitous feature of traffic jams that it has become a way—a metaphor as well as an actual thing—that Dakarois refer to all ventures in life that are subject to frustration and blockage.

22 Robert Kaplan's *The Coming Anarchy* (2000) and George Packer's *New Yorker* (2006) article on the mega city are classic statements of the apocalyptic perspective on Lagos. For a more recent iteration of this position, see the article on Lagos roads, Jeffrey Hammer, "The World's Worst Traffic Jam," *Atlantic*, July/August 2012.

23 Matthew Gandy makes this point in his article, "Learning from Lagos," when he re-counts the many exhibitions—*Century City* (2001) at the Tate Modern, the Documenta 11 (2001) in Kassel, and *Africas: Art and the City* (2002) in Barcelona—that have made Lagos a particularly dense site of discussion for contemporary urban life.

24 See also the introduction to this volume.

REFERENCES

Apter, Andrew. 2005. *The pan-African nation: Oil and the spectacle of culture in Nigeria*. Chicago: Chicago University Press.

Barad, Karen. 2007. *Meeting the universe halfway: Quantum physics and the entanglement of matter and meaning*. Durham: Duke University Press.

Bear, Laura. 2015. *Navigating austerity: Currents of debt along a south Asian river*. Stanford, CA: Stanford University Press.

Benjamin, Walter. 1999. *The arcades project*. Cambridge, MA: Belknap Press.

Bennett, Jane. 2010. *Vibrant matter: A political ecology of things*. Durham: Duke University Press.

Bize, Amiel. 2017. Jam-space and jam-time: Traffic in Nairobi. In *The making of the African road*, ed. Kurt Beck, Gabriel Klaeger, and Michael Stasik, 58–85. Leiden: Brill.

Bowker, Geoffrey, and Susan Leigh Star. 2000. *Sorting things out: Classification and its consequences*. Cambridge, MA: MIT Press.

Braun, Bruce, and Sarah J. Whatmore, eds. 2010. *Political matters: Technoscience, democracy and public life*. Minneapolis: University of Minnesota Press.

Cary, Joyce. (1939) 1989. *Mister Johnson*. New York: New Directions.

Chu, Julie Y. 2014. When infrastructures attack: The workings of disrepair in China. *American Ethnologist* 41(2): 351–367.

Coole, Diana H., and Samantha Frost, eds. 2010. *New materialisms: Ontology, agency, and politics*. Durham: Duke University Press.

Coronil, Fernando. 1997. *The magical state: Nature, money, and modernity in Venezuela*. Chicago: University of Chicago Press.

Dalakoglou, Dimitris. 2017. *The road: An ethnography of (im)mobility, space, and cross-border infrastructures in the Balkans*. Manchester, UK: Manchester University Press.

Eichenbaum, Boris. 2004. The formal method. In *Literary theory, an anthology*, ed. Julie Rivkin and Michael Ryan, 7–14. Oxford: Blackwell.

Foucault, Michel. 1972. *The archaeology of knowledge*. London: Tavistock.

Gandy, Matthew. 2005. Learning from Lagos. *New Left Review* 33: 36–52.

Giedion, Siegfried. (1928) 1995. *Building in France, building in iron, building in ferro-concrete*. Santa Monica, CA: Getty Center for the History of Art and the Humanities.

Graham, Stephen and Simon Marvin. 2001. *Splintering urbanism: Networked infrastructures, technological mobilities and the urban condition*. New York: Routledge.

Harvey, Penny and Hannah Knox. 2015. *Roads: an anthropology of infrastructure and expertise*. Ithaca, NY: Cornell University Press.

Hookway, Brandon. 2014. *Interface*. Cambridge, MA: MIT Press.

Jakobson, Roman. 1985. Closing statements: Linguistics and poetics. In *Semiotics: An introductory anthology*, ed. Robert E. Innis, 145–175. Advances in semiotics. Bloomington: Indiana University Press.

Kaplan, Robert D. 2000. *The coming anarchy: shattering the dreams of the post cold war*. New York: Random House.

Kern, Stephen. 1983. *The culture of time and space 1880–1918*. Cambridge, MA: Harvard University Press.

Khan, Naveeda. 2006. Flaws in the flow: Roads and their modernity in Pakistan. *Social Text* 24(489): 87–113.

Kittler, Friedrich. 1987. A discourse on discourse. *Stanford Literature Review* 3(1): 157–166.

Kohn, Eduardo. 2013. *How forests think: Toward an anthropology beyond the human*. Berkeley: University of California Press.

Koolhaas, Rem, Stefano Boewri, Sanford Kwinter, Nadia Tazi, and Hans Ulrich Obrist. 2000. *Mutations*. Barcelona: Actar.

Larkin, Brian. 2008. *Signal and noise: Media, infrastructure, and urban culture in Nigeria*. Durham: Duke University Press.

———. 2013. The politics and poetics of infrastructure. *Annual Review of Anthropology* 42: 327–343.

Latour, Bruno. 1993. *We have never been modern*. London: Harvester Wheatsheaf.

Lefebvre, Henri. 1991. *The production of space*. Cambridge, MA: Blackwell.

Lemke, Thomas. 2015. New materialisms: Foucault and the "government of things." *Theory, Culture and Society* 32(4): 3–25. https://doi.org/10.1177/0263276413519340.

Levine, Caroline. 2015. *Forms: Whole, rhythm, hierarchy, network*. Princeton, NJ: Princeton University Press.

Limbert, Mandana E. 2010. *In the time of oil: Piety, memory, and social life in an Omani town*. Stanford, CA: Stanford University Press.

Mackenzie, Adrian. 2002. *Transductions: Bodies and machines at speed*. London: Continuum.

Mattelart, Armand. 1996. *The invention of communication*. Translated by Susan Emanuel. Minnesota: University of Minnesota Press.

Mbembe, Achille. 2001. *On the postcolony*. Berkeley: University of California Press.

Mbembe, Achille, and Janet Roitman. 1995. Figures of the subject in times of crisis. *Public Culture* 7(2): 323–352.

Melly, Caroline. 2017. *Bottleneck: Moving, building and belonging in an African city*. Chicago: University of Chicago Press.

Moody, P. R. 1967. Road and bridge in Joyce Cary's African novels. *Bulletin of the Rocky Mountain Modern Language Association* 21(4): 145–149.

Mrázek, Rudolf. 2002. *Engineers of happy land: Technology and nationalism in a colony*. Princeton, NJ: Princeton University Press.

Ngai, Sianne. 2012. *Our aesthetic categories: Zany, cute, interesting*. Cambridge, MA: Harvard University Press.

Packer, George. 2006. The megacity. *The New Yorker*. Nov 13, 64–75.

Peters, John Durham. 2015. *The marvelous clouds: Toward a philosophy of elemental media*. Chicago: University of Chicago Press.

Rancière, Jacques. 2006. *The politics of aesthetics: The distribution of the sensible*. New York: Continuum.

——. 2009. *Aesthetics and its discontents*. Malden, MA: Polity Press.

——. 2010. *Dissensus: On politics and aesthetics*. New York: Continuum.

Shklovsky, Victor. 1965. *Art as technique*. In *Russian formalist criticism: Four essays*. Lincoln, NE: University of Nebraska Press.

Simondon, Gilbert. 1958. *On the mode of existence of technical objects*, trans. Ninan Mallamphy. Paris: Aubier, Editions Montaigne.

Star, Susan Leigh. 1999. The ethnography of infrastructure. *American Behavioral Scientist* 43(3): 377–391.

Stendhal. 1958. *The life of Henry Brulard*. London: Merlin.

von Schnitzler, Antina. 2016. *Democracy's infrastructure: Techno-politics and protest after apartheid*. Princeton, NJ: Princeton University Press.

Watts, Michael J. 1992. The shock of modernity: Petroleum, protest, and fast capitalism in an industrializing society. In *Reworking modernity: capitalisms and symbolic discontent*, by Allen Pred and Michael J. Watts, 21–64. New Brunswick, NJ: Rutgers University Press.

West, Matthew Ellis. 2015. Intellectual property and the knowledge economy's global division of labor: Producing Taiwanese green-technology between the United States and China. Ph.D. dissertation, Columbia University.

Williams, Raymond. 1977. *Marxism and literature*. Oxford: Oxford University Press.

8

Sustainable Knowledge Infrastructures

GEOFFREY C. BOWKER

Prolegomenon

In the introduction to this volume, Akhil Gupta, Hannah Appel, and Nikhil Anand explore (albeit somewhat negatively) the promise of infrastructure (I shall return to this). But before I get to the nature of that promise, I will talk about whether knowledge infrastructures fit with the other infrastructures of this volume.[1]

My starting point is an observation by Michel Serres (1993) that while the Roman empire produced long-lasting infrastructure (roads, aqueducts), the Greek empire gave us Euclid, whose geometry still structures so much of the way in which we think. So are aqueducts and systems of geometry the same sort of thing? I will take as given that infrastructures are always relational: one person's infrastructure is another's site (Star and Ruhleder 1996). So to be infrastructural is to be in a subtending relationship with.

Taking this lightweight definition, it is clear that the taken-for-granted nature of turning on a tap and expecting water to come out of the faucet is equivalent to turning to an accredited journal and expecting knowledge to come out. Just as the Environmental Protection Agency has water-quality regulations, so does the academic literature in theory set parameters around knowledge quality. Similarly, when I read that the work in this volume "shows how oil rigs and electrical wires, roads and water pipes, bridges and payment systems articulate social relations to make a variety of social, institutional, and material things (im)possible" (Appel, Anand, and Gupta, this volume), I feel no doubt that knowledge infrastructures do similar work. The question is not so much "what is an infrastructure?" as "when is an infrastructure?": When is it useful to use the term to connect across an array of literature (cf. Engestrom 1990)?

However, there is a difference between a marginally robust analogy and an equivalence—the latter is my preferred space for this chapter. Timothy Mitchell (2011) posits a direct relationship between the form of energy that we consume and the social (and, I would argue, religious) theories we produce. This is highly unsurprising in a sense—a society that seeks to pillage past sunlight stored over many millions of years within several generations while knowing the very limitations of that resource is making a series of strong statements:

- That our generation is the ultimate or penultimate generation—so we need to maximize for resources in the present rather than distribute them in the future. This form of argumentation is highly pervasive—so much biodiversity rhetoric is about preserving maximal biodiversity for our usage now, rather than maximizing for life's ability to generate new forms in the future. This apocalyptic vision preceded the second industrial revolution (late eighteenth century) in the form of millennialism—what is interesting is that the apocalypse jumped from Christianity to rational science with nary a leap of faith in sight. So, après nous le déluge. . . . Spend, spend, spend (Nicholson and Smith, 1977).
- That energy is a finite resource to be exploited rather than a sustainable resource to be nurtured. Late capitalism has precisely configured workers as such an exploitable resource—the language of energy has pervaded our social fabric.
- That we are the advancing edge of the future. The very discourse of certainty and optimism in progress attended the development of the steam engine and the massive exploitation of coal as a resource. We could "accelerate" human progress by exploiting the ship and the train run by steam. This generation, it was felt, was different from any previous generation because it could simultaneously speed things up and annihilate distance. The *status quo ante* only really worked when energy was an infinite resource.

But what has this to do with knowledge practices? Let me evoke the resource of *Coins, Bodies, Games, and Gold* by Leslie Kurke (1999). She deploys a poetic semiotics to find out how people were talking about money for several centuries in Greece after its instauration and before Aristotle's philosophical treatise (about three hundred years). She argues that such an overwhelming social fact as this invention could not have been ignored—and that it is through semiotic analysis that we can uncover how this discussion occurred without the abstract concept being used. Similarly with energy—much of our knowledge discourse is about it, we just need to understand how to read that discourse, with all its flows, its stocks of intellectual capital, and its circulation . . . not to mention its currency.

Ontologically, it seems to be always in our culture a question of which came first—and I wish it were not. It is not that energy discourse undergirds epiphe-nomenal philosophical and political discourse (though this argument can be a useful propaedeutic to the question of the source of our practices). Rather, the exploitation and the discourse occur together in a single unit—there is no universal arrow of causality that leads from one to the other. I think here of Alfred Sohn-Rethel's (1975) argument, "Science as Alienated Consciousness": I have no difficulty following him and arguing that universal time and space are a meditation on the commodity form of early capitalism, providing one inverts in the same breath and argues that the commodity form is a meditation on the nature of space and time. I agree with Mitchell that we could not have the great infrastructure projects of the twentieth century without either carbon and oil or the economic theories of Marx and Keynes. We think through how we act socially and economically. So when I talk about knowledge infrastructures here, I do so in terms of a formal equivalence between the ways we are in the world (through our infrastructure) and the ways in which we think about the world. This for me is what the "determination in the last instance," referenced in the introduction, is about. Similarly, when I read in the introduction that "capitalism can be performative only because of the many means of producing stable repetition," I cannot but think of the stabilities of the modern Academy.

Introduction

Sustainability is a word in vogue right now—we want sustainable develop-ment, a sustainable relationship with nature, and, in general, sustainability. It is unclear how we can make the academic system sustainable. There are too many producers of knowledge and too few consumers: Katherine Hayles (2012), for example, notes that some 93 percent of papers in the humanities do not get cited after five years of publication—the situation is not greatly different in the social and natural sciences (the latter having to deal with the breaking apart of theoretical arguments into the least publishable units). At the same time, we are increasingly being drawn into a quantifiable, practical world (compare Shinn [1980] for this move at an earlier epoch) whose academic output, fo-cused on a behaviorist ideology of stimulus and response, is not sufficient to the issues of our times—where only genuinely transdisciplinary thinking can address the issue of both surviving and thriving in the Anthropocene.

There are two sides to this equation. How do we reimagine the nature of knowledge for the way the world is now? How do we put into place forms of knowledge production that can bear the weight of these new exigencies?

Before We Begin

We are still, in the so-called developed world, locked into a knowledge production system inherited from the Enlightenment. From that epoch, we have our current classification of knowledge—soft/hard sciences, the humanities versus the sciences, the organization of knowledge into "disciplines" (for all that we recognize in practice that these are ridiculous divides). We have inherited the great classifications of knowledge, from Auguste Comte (1830), Ampère (1834), and others, who tell us that the world of knowledge can be shoehorned into a set of *a priori* categories—physics tells about the material world in raw form; chemistry about when the physical forms start to interact; biology about when those interactions become sufficiently complex as to self-organize; and sociology about when self-organizing biological units achieve awareness. As you walk through many august campuses, you can see this classification inscribed in stone. When I worked at the University of Illinois at Urbana-Champaign, we used to talk about the north and south ends of campus. Appropriately to the metaphor, the north end was science and engineering, the south end was humanities and social sciences, and somewhere off in the deep south was agriculture and veterinary science. In my current circular campus—the University of California at Irvine—there is a void in the middle, so I am forced to walk to my left from information and computer science to meet engineering and then the hard sciences, or to my right and meet first anthropology and the social sciences and finally the humanities and the arts (which are over a bridge, duly removed from the circle). These divides are not just notional; as they get inscribed into the landscape they shape who meets whom with what regularity—they order the possibility of discourse.

New Forms of Knowledge Expression Require New Knowledge Infrastructures

All academic fields include some array of narratives, numbers, visualizations, computer code and files, physical prototypes, databases, organic and inorganic materials, and some forms of knowledge-bearing objects, each with particular materialities. That the written record is a standardized and normalized form of academic communication is unquestionable. However, the portfolio of artifacts generated by scholars suggests that universally reducing forms of knowledge to linear text may, in many cases, be a vestigial or anachronistic activity persisting through an inertial scholarly infrastructure. (A general lack of community standards to judge the scientific rigor of newly emerging objects provides potential evidence of such inertia—though see Sousanis [2015] for a wonderful counterpoint.) The broad epistemic cultures of the academy all have their own historical "ways of knowing" ratified and internalized through sociotechnical processes. New forms of knowledge give rise to novel representations, artifacts, and objects that reflect significantly new ways of thinking about phenomena. Meaningful academic work is not the sole domain of the traditional laboratory scientist or library-bound scholar; significant research and knowledge work exists at all levels of the academic process, and it often goes unrecognized or unacknowledged. Scholars and institutions widely recognize that knowledge communication is changing and the forms of reportable scholarship are proliferating at a high rate, frequently citing the need for evolution within the scientific communities and the academy to acknowledge and reward emergent ways of knowing.

However, the work of establishing new evaluation schemes, metrics, and grounded standards of intellectual merit is misaligned with the scientific and professional demands of most academics. Thus, the work of producing mechanisms and scholarship to recognize and render normative transformative research either lays fallow or proceeds at a pace outstripped by the need for and realization of innovation.

At its base, the performance of contemporary scholarship demands interdisciplinary skill; however, the expectations and training of knowledge workers implicitly focus on the domain alone. Interdisciplinary projects, particularly those headed by a single investigator, demand trained interdisciplinarians to conduct them. The objects of study, as well as the informing theoretical bases, require the triangulation and synthesis of multiple methodologies, both qualitative and quantitative, and call upon the ability of investigators to integrate multiple epistemic viewpoints—operating in what Peter Galison (1997) refers to as "trading zones," working through and across the boundaries that traditionally separate distinct fields of research.

There is no simple alignment between historically defined disciplines and the knowledges communicated via recently emerged and currently emerging scholarly forms. The continued sustainability of scholarship is predicated on a deeper understanding of such an alignment. We need to first identify and compare new forms of knowledge expression and the teams that produce them; and then we must characterize in de-disciplinary terms the knowledge that they carry. Within traditional disciplinary boundaries, there is a relatively stable and limited set of research methods, major questions, and recognized, rewarded, acknowledged knowledge artifacts. The bounding of discipline into methodological categories is already quickly becoming obsolete as much work is to some extent multimethod. As we explore knowledge artifacts outside of these traditional boundaries, the knowledge itself—the outcome of the research—becomes the boundary object that brings together the efforts of previously disparate scholarly communities.

We cannot look at new products without also investigating the ways in which the richness of knowledge itself is changing. Since the early nineteenth century, during which scholarship was segregated into disciplines, richness was achieved through the complementary presence of artifact and researcher, the movement of knowledge workers through institutions, and the insertion of the artifact into the archive. There is no reason that such a richness cannot be contained, even heightened, in the present by means of multimodal scholarly artifacts. The appearance of new forms of knowledge expression is an opportunity to reveal aspects of knowledge that have remained absent or underspecified in the past.

I propose a strategy for a new integration of the humanist sciences, social sciences, and physical/natural sciences. First, I propose a "reasoned ecological survey" of new modes of knowledge expression. Second, I argue the need for a new alliance between the traditionally conceived humanities and the social and natural sciences, and I sketch out a pathway to its development.

In his *Social History of Knowledge*, Peter Burke (2012: 162) writes, "It is well known that the word 'scientist' in English is a coinage of the 1830s." According to the *Oxford English Dictionary*, in 1834 the *Quarterly Review* described the first use of the term "scientist" as a means to unify fields of inquiry generally referred to as the sciences: "Science . . . loses all traces of unity. A curious illustration of this result may be observed in the want of any name by which we can designate the students of the knowledge of the material world" (*OED*, "scientist" entry). Given that to classify is human (Bowker and Star 2005), it seems reasonable that the British Association for the Advancement of Science desired such a lexical item, the referent of which neatly comprises the general population of research practitioners concerned with the material world.

Knowledge classification is a double-edged sword with which the fruits of both physical and conceptual worlds are sliced into heterogeneous sections not easily combined. It supports and reinforces the conceptual differences between entities. Categories deemed "unlike" are likely to remain divorced through the classificatory reification of their differences. For example, scientists in the 1830s all became "students of the knowledge of the material world," and students who did not fit into such a category necessarily became something else. They filled the negative conceptual space of science, the unmarked category. The force with which the term "scientist" was created and would continue to be wielded through to the current day meant that all nonscientists belong to the category of the humanities. (Interestingly, the term "humanities" had already been in existence for at least some 350 years at the coinage of "scientist.") Our contemporary infrastructure for knowledge production has only recently begun to evince the necessity of functionally reconceptualizing and reclassifying the relationship between the humanities and the sciences.

Following the rise of specialization in academia beginning at the end of the Industrial Revolution, the extent to which the humanities and the sciences interact and support one another has become increasingly unclear. The humanities/sciences dichotomy evinced by C. P. Snow (1964) evidences a general perception of their relative lack of formal knowledge-productive interaction. However, it is no longer productive to conceptualize academia solely as a loose constellation of disciplinary silos, each belonging exclusively to either the humanist or scientific camp. Indeed, with Bruno Latour's (2003) questioning of the nature of modernism through a critique of its underlying dichotomy between the natural and the social, academia might now be considered to exist in a period of radical classificatory change. The rise of the term "sociotechnical" (first appearing in 1920) in scholarly investigations implies an inherent relationship between the social (human) and the technical (scientific).

Although scientists have consistently found new ways of developing new tools for detection, analysis, and communication; creating databases and digital repositories or accessible archives; and representing knowledge over the past forty years of digital scholarship, we do not have a generalized understanding of how to recognize these new forms of knowledge explicitly, how to understand what makes them work. While there is a general recognition of the need for high-risk/high-reward research activities, it is hard to sculpt a career performing them. We are moving more and more into the world of soft money in academia and to ever more stringent accounting regimes that cherish rate of production.

Thomas Kuhn's *The Structure of Scientific Revolutions* (1962) details the difference between normal science and revolutionary (or transformative) science. The distinction between an incremental science, which supports current practices and theories, and a science that lays radically new pathways to understanding, remains salient. Scientific epistemologies become deeply embedded and difficult to challenge openly because they, like other technologies, become engrained in infrastructures. Reticence to consider new forms of expression, new ways of knowing, and objects outside of the standards of a scientific community impedes the mechanisms of transformation. It takes time, strategy, and a realignment of incentives to bring scientific communities to a state of openness where new epistemologies and ontologies can be considered openly. It is no easy task to balance between rigorous critique and hopeful compassion, working to break down the logjam of doubts about the relationship between novel forms and their ability to bear knowledge.

The current practice of scientific communication narrows the band of transmission to a channel dominated by linear text. The explosion of tools and representations through which scientific thought is being expressed suggests a collective desire to engage multiple channels and modalities of knowledge transmission. There is a potential to create legitimate conduits for recognizing and valuing knowledge expressions arising outside of the traditional institutional boundaries, as is the case of intellectually worthwhile activities found in citizen science and the maker and DIY communities (cf. Ratto 2011).

Designing New Knowledge Infrastructures

A brief developmental description of scholarship's current infrastructural configuration begins in the long eighteenth century—knowledge organization at the time of the Enlightenment ushered in the epoch of "x-ologies" (Serres 1989), during which knowledge practices could be classified into disciplines that could be entrenched in institutions. It is in the contemporary tail end of this nested epoch in which much of contemporary scholarship resides. But the general rise of "x-ologies" depended on a particular type of system to support it. As William Warner and Clifford Siskin (2010) argue, the Enlightenment made good on Francis Bacon's goal of turning science into a factory system, which could produce progressive knowledge through the principle of the division of labor. There were three key components for making this work. First, one needed a single universal classification of knowledge, which would allow each academic laborer to work in a well-defined, specialized area that would not impinge on its neighbors. Comte and Ampère rose to the challenge. Knowledge had to be produced within a consensus ontology that cut nature at the joints just so. Second, one needed a form of scholarly communication appropriate to the task—each discipline would have its major journals, occasionally spinning off subjournals representing subdisciplines—but never folding back into a questioning of the ontology. Third, and this is perhaps least obvious, it needed a scientific community of the order of that obtaining in Europe in the early nineteenth century—numbering in the thousands rather than the millions of today. This latter need revolved around the issue of readership—to know a field one had to read the major journals therein, which became increasingly untenable as academic production multiplied. So the system coevolved along three axes—ontology (splitting the world into mutually exclusive types of being), communication (having a consistent means of communication, which would enable the progress of the resultant disciplines), and community (having the academic labor force produce knowledge at a pace at which it could be reasonably consumed).

The need for new, integrative knowledge forms has slowly become more apparent. Over the past two hundred years—broadly since Malthus, and summarized nicely in Heidegger's concept of "standing reserve"—we have recognized ourselves as planetary managers: we garner a huge percentage of the world's natural resources; we manage water, plants, landscapes, and so on. Tackling these issues has involved creating interdisciplinary knowledge. Vannevar Bush (1947) made this a central call in the founding of the National Science Foundation. Thus, interdisciplines have proliferated wildly since the 1940s: problem-based science (Mode II science; Nowotny et al. 2001) necessarily involves complex lateral communications between multiple disciplines.

We can no longer imagine a world in which each separate discipline could work in isolation and report its products back up the chain so that a set of philosophers could adduce a final synthesis: the work of synthesis must be local to a community, pragmatic, and integral to daily knowledge practice. No one ontology will suffice—some of the most interesting recent scholarship crossing the traditional divides addresses the development of the Anthropocene (human-dominated) geological era, blending insights from artists, humanities scholars, and social and natural scientists. The concept of the Anthropocene, which Peter de Bolla (2013) would call a "load bearing concept," does not recognize the siloed ontologies of our traditional academic divisions. Accordingly, we need new bundles of technology, vocation, narration, organization, institution, analysis, and communication that we can use to theorize our current infrastructures of scholarly and knowledge communication and participation. Only through changing scholarly communication can we shift the currents and eddies of power that will allow us to design a sustainable knowledge infrastructure for our species.

Broadly speaking, we need to recognize and train new types of knowledge workers (brokers and transducers), who can move from atomistic to holistic knowledge forms, and to create appropriate new knowledge-bearing forms. Knowledge brokers, as a functional role within an academic ecology, move among multiple communities and are adept at making the connections across social and scholarly networks necessary to spur creative knowledge work. Brokers foster cross-disciplinary innovation by bringing together scholars and researchers who would otherwise be isolated from each other, providing the seed of a fruitful conversation, then mobilizing that conversation into a new kind of research collaboration. Another critical category within this ecology are knowledge transducers (a metaphor borrowed from electrical engineering) possessing the skills to transform data, knowledge, and practice in one arena and prepare it for effective use in another. Through exploring new research networks we locate the knowledge and practice of knowledge brokers and transducers working across the borders of traditional disciplines. In working to change the way we think about knowledge and its communication, brokers and transducers make existing cross-disciplinary work visible and amenable to structures of professional recognition, possibly revealing a new core set of skills, competencies, and perspectives necessary for contemporary scholarship.

We are in the first period of being able to theorize infrastructures as they develop. The first appearance of the term "infrastructure" has been traced to 1927 (with a brief but lonely appearance in late 1880s France) to describe the complex network of roadways, waterways, and communication systems facilitating military mobilization in the United States. In less than a century, a formative intellectual market has arisen to highlight and render visible many historical projects that have grown, agglomerated, and now invisibly serve as carriers of heterogeneous resources and providers of necessary services and functions. Waterways, electrical grids, roadways, oil pipelines, and large-scale networked computing and communications are common objects of infrastructural discussion; however, the power of infrastructural thought lies in understanding the hybrid social and technical natures and histories of such projects. We are finding our way not only to describe and understand the dynamic nature of existing infrastructures, but also to attempt to foster and build new ones to suit our needs. This set of activities is unprecedented in human history—to design at a level of scale far beyond human capacity for monitoring, control, or predicted use. An analogy here might be with the practice of classification in the nineteenth century, which has been termed the century of classification. While there were combined and uneven developments of genetic classification systems (those based on the origin or seed of the object under study [Tort 1999]), there was no generic science of classification.

Often, our awareness of infrastructure arises at its strain or breakdown. This is becoming more and more apparent with the rise of technological forms that challenge the existing knowledge infrastructure based on logocentric artifacts. The traditional scholarly journal article, monograph, or conference paper re-creates and re-inscribes the inertia of eighteenth- and nineteenth-century technologies, resistant to new modes of expression. It requires potential new expressive forms of knowledge to be rendered down into a purely textual format for dissemination, and entry into both scholarly canons as well as established reward systems misses both the point of progress, as well as the opportunity for the infrastructures of knowledge and scholarship to keep apace of contemporary transformations in communication.

And yet scholars and builders consistently find new ways to display their intellectual work. As many fields (re)discover fusions with design practices—both virtual and material—a wide array of knowledge-bearing objects, systems, and environments emerges. As they enter into a system for which scholars must be rewarded for their efforts, contributions, and time, the inertial system is slow to develop communally established hermeneutics of the digital, and heuristics for evaluating their contributions.

Linear text is very good for creating, in Latour's (1987) term, immutable mobiles: the feature of written works to be durable through space and time, imbuing them with a sense of authority, is a deeply held and foundational value in the work of scholarship. Still, as Alfred North Whitehead points out, this type of concretism conflates the form with the embedded knowledge and commits a fallacy of reification (Stengers 2011). Picking up on this thread with respect to information technologies, Susan Leigh Star ([1995] 2016) unpacks the concept of misplaced concretism in which we are rendered beholden to the technological infrastructures through which our knowledge is stored, disseminated, and consumed. As the concretized forms of scholarship—journal articles, monographs, and conference papers—developed the sine qua non of knowledge cultures, logocentric technologies become clearer. We are indeed far from the days of Socrates's argument that philosophy should not be written down (ironically transmitted through the writings of Plato [Kurke 2011]). Contemporary practices of scholarship are far broader in terms of expressive forms. Challenging the reification of the written word may seem heretical in the short-term context of eighteenth- and twentieth-century scholarly communication; however, the broader swath of historical analysis (including Bender and Marrinan's [2004] wonderful analysis of plates and diagrams and Tufte's [1983] work on historical visualization) suggests that nonlexical representations have enduring histories of their own, despite the fact that they have not maintained the centrality and inscrutability achieved by text.

We must give durable and transportable form to our ideas to create knowledge cultures, archives, histories, and practices. Materiality is a necessary condition for the transference and proliferation of knowledge, whether digital or analog. We need a new, apposite, materiality.

By repositioning the discourses of materiality, which are most often interpretive, as active spaces for the design of scholarly objects, artifacts, and expressions, we confront the opaqueness of reified forms. Folding theories of design into the more traditional activities of classification and taxonomy sets up the project as an open and extensible program.

To deploy a definition of scholarly communicative platforms from a 2011 Harvard and Microsoft research conference, "Transforming Scholarly Communication," "a platform is more than just 'software': it is in fact an ecosystem that includes software, data, services and people. It is in its essence sociotechnical, and its function is to enable research and scholarly communication" (Abbot et al.). Instances, in general, are the singular, stand-alone outputs derived from the productive use of a given platform. As an example, one might consider the (sadly defunct) digital humanities e-journal *Vectors*. This multimedia journal is clearly reliant upon software to bring its digital existence to life, but it is similarly reliant upon a community of scholars and programmers for the creation and dissemination (service) of its content (data). Within the *Vectors* platform, one finds dozens of individual instances, or scholarly artifacts, in a form facilitated (perhaps uniquely) through the use of the *Vectors* platform. A notable example can be found in "Blue Velvet," presented in the platform's "Difference" issue (Goldberg and Hristova 2007).

The idea of the platform is itself a politically weighted concept that groans under an overabundance of meaning and connotation in political, technical, and social spaces (Gillespie 2010). The platform, in Tarleton Gillespie's analysis, is a place from which to speak, a source of opportunity, a political goal, and a technological grounding. The platforms we are concerned with are newly standardized to a local set of instances of expression that may be but are not necessarily extant as an artifact of that expression. As we identify some new mode of knowledge representation or production, we also identify the network of relationships in which it exists, and of particular interest is the way in which that instance of knowledge relates to academic, technical, and other standards. In the confluence and stabilization of the relationships is both the grounding and basis for new platforms formed of and being a stable expression of a set of standards relationships consistent to a given set of instances of knowledge expression.

The platform stabilizes the standards while paradoxically emerging from their stability and relationship in expression. To be a standard, something must enable interoperability, establish minimal levels of quality, or reduce variation across a given range of objects.

Standards may also be de facto, or emergent, as well as *institutional*, or produced by the coordinated effort of some authoritative group. An expressive platform such as YouTube (an example that is nonacademic in origin) required the stable expression of a variety of connectivity (TCP/IP, HTML, video streaming) and physical (wired and wireless infrastructure), technical (video compression, sound compression), and content standards. Without the stability of those standards in use, the platform does not function, but the making of short, early YouTube-esque videos did not depend on the existence of the platform; rather the stabilization of the standards set allowed for the growth, wide access, and spread of the expressive instances linked to that platform.

I have worked with a wider range of scholars on the Re:Enlightenment project, whose humble aim is to create new knowledge forms and associated knowledge infrastructures appropriate to our current epoch. A groundbreaking book, Cliff Siskin and William Warner's edited collection *This Is Enlightenment* (2010), forged a new understanding of the Enlightenment centered on forms of knowledge expression and communication, such as the Encyclopédie, enabled by new information and communications technology. Through a series of international meetings, we have brought leading scholars across the humanities together with theorists of new media in order to theorize old forms of scholarly communication and to imagine new forms. For example, the Royal Society for the Arts in London has espoused the movement and has developed an extremely rich new form of knowledge visualization through RSA Animate.[2]

Our current hypothesis is that knowledge is "stuck" today and that new forms of "de-disciplinary" collaboration and new forms of knowledge expression must develop together. By "stuck," I do not mean that progress has ceased within the received disciplinary frames. Valuable knowledge continues to be produced. But a key marker of knowledge being stuck is that progress can get in the way of more consequential advances. So too can good intentions—as in the interdisciplinary celebration of difference and sharing. That gesture of mixing only conserves "discipline" itself as the shape of knowledge, dispersing the energy of innovation into the now hard-wired circuits of the platform we call disciplinarity.

The need to change platforms is thus our takeoff point—a point similar in kind to Bacon observing that Scholastic knowledge was "stalled" in the early seventeenth century, requiring a collaborative effort to restart it.

- Being stuck in this manner is not a historically frequent or a local problem. It is rare and consequential, for it is experienced across the entire organization of knowledge.
- Solving the problem entails some recognition of habits that do not work anymore (Bacon's "idols") and some hard decisions about what should stop. But most of all it is about figuring out how to take steps

forward. That is where optimism—rather than crisis, worry, and blame—come into the equation. Bacon insisted that this was not a matter of his predecessors being wrong or not as smart, but of having the "good fortune" of living at a moment when limitations become visible and new "resources" raise the possibility of working in a new way.

- Living in our own moment of new resources—especially the algorithmic and the electronic—raises in every discipline and other kinds of knowledge groupings the problem of how to scale up to the possibilities they open. Bacon's answers included new methods (method is, etymologically and as used by Bacon, "a way forward"), new genres (aphorism and essay), and new agendas (his list of 130 histories to be ticked off first, together with new collaborative tactics, led to the creation of the Royal Society).

This challenge is of more than intellectual concern. The institutions in which most knowledge workers live and labor have not kept pace, or have done so piecemeal, without a long-term vision or a strategy. For example, the widespread excitement about crowdsourced knowledge, assembled by unpaid individuals who volunteer their time out of personal interest, ignores the fact that most knowledge workers' salaries are still paid by brick-and-mortar organizations with hierarchical structures, established institutional cultures, systems of credit and compensation, and other "sticky" processes and routines. Similarly, our educational systems, libraries, publishers, news organizations, intellectual property structures, and political mechanisms have struggled to match or adapt to the changing information environment (Borgman 2007). The result is a patchwork of unsatisfactory kludges, contradictions, and inconsistencies that may undermine the prospects for change.

Popular attention and academic research on changing knowledge systems have tended to follow the new, fast-moving, and dramatic parts of the current transition. For example, in *Reinventing Discovery*, Michael Nielsen (2012) extrapolates from current events to the eventual rise of a scientific culture of "extreme openness" where "all information of scientific value, from raw experimental data and computer code to all the questions, ideas, folk knowledge, and speculations that are currently locked up inside the heads of individual scientists" is moved on to the network, "in forms that are not just human-readable, but also machine-readable, as part of a data web." Clay Shirky (2010) argues that a "cognitive surplus" will permit massively distributed contributions to the analysis of information and the production of new knowledge. While surely partially correct, these breathless assessments too often lose track of crucial questions about the complex processes of mutual adjustment by which older knowledge institutions adapt to emergent ones, and vice versa. Charmed by the novelty of the first date, they miss the complexity of the marriage that ensues: the dynamics of scale, time, and adjustment by which new practices emerge.

Given the layered nature of infrastructure, navigating among different scales—whether of time and space, of human collectivities, or of data—represents a critical challenge for the design, use, and maintenance of robust knowledge infrastructures. A single knowledge infrastructure must often track and support fluid and potentially competing or contradictory notions of knowledge. Often invisible, these notions are embodied in the practices, policies, and values embraced by individuals, technical systems, and institutions. For example, sustainable knowledge infrastructures must somehow provide for the long-term preservation and conservation of data, of knowledge, and of practices (Borgman 2007; Ribes and Finholt 2009). In the current transformation, sustaining knowledge requires not only resource streams, but also conceptual innovation and practical implementation. Both historical and contemporary studies are needed to investigate how knowledge infrastructures form and change, how they break or obsolesce, and what factors help them flourish and endure.

A quintessential tension surrounds the deployment of standards and ontologies in knowledge infrastructures. Fundamentally, it consists in the opposition between the desire for universality and the need for change. Robust hypotheses require information in standardized formats. Thus, the spread of a particular disease around the world cannot be tracked unless everyone is calling it the same thing. At the same time, medical researchers frequently designate new diseases, thus unsettling the existing order. For example, epidemiologists have sought to track the phenomenon of AIDS to periods predating its formal naming in the 1980s (Grmek 1990). However, using historical medical records to do so has proven difficult because prior record-keeping standards required the specification of a single cause of death, precluding recognition of the more complex constellation of conditions that characterize diseases such as AIDS.

How might one solve this problem (if it is solvable at all)? One could review the old records and try to conjure them into modern forms. This could work to an extent; some fields, such as climate science, routinely investigate historical data before adjusting and restandardizing them in modern forms to deepen knowledge of past climates (Edwards 2013). Yet this is possible largely because the number and variety of records are relatively limited. In many other fields such a procedure would be extremely difficult and prohibitively expensive. Alternatively, one could introduce a new classificatory principle, such as the

Read Clinical Classification, which would not permit that kind of error to propagate. Here too, due to the massive inertia of the installed base, it would cost billions of dollars to make the changeover—this is why Ted Nelson's better structure for the web will never happen. On top of that, it would complicate backward compatibility: every new archival form challenges the old (Derrida 1996). In practice, this adds up to a very slow updating of classification standards and ontologies, marked by occasional tectonic shifts.

Today, hopes for massively distributed knowledge infrastructures operating across multiple disciplines consistently run headlong into this problem. Such infrastructures are vital to solving key issues of our day: effective action on biodiversity loss or climate change depends on sharing databases among disciplines with different, often incompatible ontologies. If the world actually corresponded to the hopeful vision of data-sharing proponents, one could simply treat each discipline's outputs as an "object" in an object-oriented database (to use a computing analogy). Discipline X could simply plug discipline Y's outputs into its own inputs. One could thus capitalize on the virtues of object orientation: it would not matter what changed within the discipline, because the outputs would always be the same. Unfortunately, this is unlikely—perhaps even impossible—for both theoretical and practical reasons. An "object-oriented" solution for these incompatibilities is theoretically improbable because the fundamental ontologies of disciplines often change as those disciplines evolve. This is among the oldest results in the history of science: Kuhn's term "incommensurability" marks the fact that "mass" in Newtonian physics means something fundamentally different from "mass" in Einsteinian physics (Kuhn 1962). If Kuhnian incommensurability complicates individual disciplines, it has even larger impacts across disciplines. A crisis shook virology, for example, in the 1960s when it was discovered that "plant virus" and "animal virus" were not mutually exclusive categories. Evolutionary biology suffered a similar, and related, crisis when it was learned that some genes could jump between species within a given genus, and even between species of different genera (Bowker 2005). Suddenly, disciplines that previously had no need to communicate with each other found that they had to do so, which then required them to adjust both their classification standards and their underlying ontologies.

In practice, an object-oriented solution to ontological incompatibilities is unlikely because we have not yet developed a cadre of metadata workers who could effectively address the issues, and we have not yet fully faced the implications of the basic infrastructural problem of maintenance. We do know that it takes enormous work to shift a database from one medium to another, let alone to adjust its outputs and algorithms so that it can remain useful both to its home discipline and to neighboring ones. Thus three results of today's scramble to post every available scrap of data online are, first, a plethora of "dirty" data, whose quality may be impossible for other investigators to evaluate; second, weak or nonexistent guarantees of long-term persistence for many data sources; and, finally, inconsistent metadata practices that may render the reuse of data impossible—despite their intent to do the opposite.

We expect our knowledge infrastructures to permit effective action in the world; this is the whole impulse behind Pasteur's Quadrant or Mode II science. And yet, in general, scientific knowledge infrastructures have not been crafted in such a way as to make this easy. What policymakers need and what scientists find interesting are often too different—or, to put it another way, a yawning gap of ontology and standards separates the two. Consider biodiversity knowledge. In a complex series of overlapping and contradictory efforts, taxonomists have been trying to produce accounts of how species are distributed over the earth. However, the species database of the Global Biodiversity Information Facility, which attempts to federate the various efforts and is explicitly intended for policy use, does not produce policy-relevant outputs (Slota and Bowker in prep.).The maps of distribution are not tied to topography (necessary to consider alternative proposals such as protecting hotspots or creating corridors), they give single observations (where what is needed is multiple observations over time, so one can see trends), and for political reasons, they do not cover many parts of the planet (which one needs in order to make effective global decisions). Similarly, in the case of climate change, for decades the focus on "global climate"—an abstraction relevant for science, but not for everyday life—has shaped political discourse in ways that conflicted with the local, regional, and national knowledge and concerns that matter most for virtually all social and political units. Climate knowledge infrastructures have been built to produce global knowledge, whereas the climate knowledge most needed for policymaking is regional, culturally specific, and focused on adaptation.

Conclusion

In a prescient piece, Walter Benjamin argued that knowledge today was basically *about* databases: "And even today, as the current scientific method teaches us, the book is an archaic intermediate between two different card index systems. For everything substantial is found in the slip box of the researcher who wrote it and the scholar who studies in it, assimilated into its own card index" (cited in Gitelman 2013). If you take the written text—the canonical scientific paper—as epiphenomenal, then you open the door to rich, new expressive forms.

However, in order to do so, you need to reorganize the nature of knowledge production—the "knowledge infrastructures" that we have so painstakingly built around the word. One ideal working unit might consist, say, of biologists, database designers, graphic designers, and humanists. And yet our reward structures favor the front end of the written text—not the back end of the "slip box," which is where the most interesting work is done. We are indeed in the epoch of the database—it is the site of both useful knowledge and new modes of expression.

This brings me back to the promised promise of infrastructure from my prolegomenon. Infrastructure is so often seen as a dead weight—it is not as if I could decide to turn off all my grids: or rather if I did, I would render myself ineffective at the same time. The promise has to be as rhizomatic as the infrastructure itself is turning out to be (I am thinking here of distributed sources of solar and wind energy rather than singular power plants). Through our forms of infrastructure, we can reconsider the nature of our knowledge; and through our forms of knowledge, we can reconsider the nature of our infrastructure.

NOTES

My deep thanks to the EVoKE lab members, especially Cory Knobel, for working out an early framing of these ideas; and to the School of Academic Research group for providing such generous and creative interlocutors.

1 The triple introduction here is an homage to Jacques Derrida's *Archive Fever*, which is nothing but a set of exergues, introductions, and prolegomena—we cannot get into a subject like infrastructure without approaching it from multiple angles simultaneously.

2 RSA Animate, https://www.thersa.org/discover/videos/rsa-animate.

REFERENCES

Abbot, Mark, Taliesin Beynon, Rachel Bruce, Derick Campbell, Tim Clark, Tom Cramer, Dave De Roure, Cory Knobel, Alberto Pepe, and Thomas Robitaille. 2011. Platforms for scholarly communication. Harvard University and Microsoft Research, *Transforming Scholarly Communication* (blog). October 25, 2011. http://msrworkshop-blog.tumblr.com/post/11911100535.

Ampère, A.-M. 1834. *Essai sur la philosophie des sciences, ou Exposition analytique d'une classification naturelle de toutes les connaissances humaines*. Paris: Bachelier.

Bender, J. B., and M. Marrinan. 2004. *The culture of diagram*. Stanford, CA: Stanford University Press.

Borgman, C. L. 2007. *Scholarship in the digital age: Information, infrastructure, and the Internet*. Cambridge, MA: MIT Press.

Bowker, Geoffrey C. 2005. *Memory practices in the sciences*. Cambridge, MA: MIT Press.

Bowker, Geoffrey C., and Susan Leigh Star. 2005. *Sorting things out: Classification and its consequences.* Cambridge, MA: MIT Press.

Bowker, Geoffrey C., Stefan Timmermans, Adele Clarke, and Ellen Balka, eds. 2016. *Boundary objects and beyond.* Cambridge, MA: MIT Press.

Burke, P. 2012. *A social history of knowledge, II.* Cambridge, UK: Polity.

Bush, Vannevar. 1947. *Science, the endless frontier: A report to the president on a program for postwar scientific research.* Washington, DC: National Science Foundation.

Comte, Auguste. 1830. *Cours de philosophie positive.* Paris: Bachelier.

de Bolla, P. 2013. *The architecture of concepts: The historical formation of human rights.* http://librarytitles.ebrary.com/id/10769548.

Derrida, Jacques. 1996. *Archive fever: A Freudian impression.* Chicago: University of Chicago Press.

Edwards, Paul N. 2013. *A vast machine: Computer models, climate data, and the politics of global warming.* Cambridge, MA: MIT Press.

Engestrom, Yrjo. 1990. *Learning, working and imagining.* Helsinki: Orienta Konsolit Oy.

Galison, P. 1997. *Image and logic: A material culture of microphysics.* Chicago: University of Chicago Press.

Gillespie, Tarleton. 2010. The politics of "platforms." *New Media and Society* 12(3): 347–364.

Gitelman, Lisa, ed. 2013. *"Raw data" is an oxymoron.* Cambridge, MA: MIT Press.

Goldberg, David Theo, and Stefka Hristova. 2007. Blue velvet. *Vectors* 3(1). http://vectors.usc.edu/issues/5/bluevelvet.

Grmek, M. D. 1990. *History of AIDS: Emergence and origin of a modern pandemic.* Princeton, NJ: Princeton University Press.

Hayles, Katherine. 2012. *How we think: Digital media and contemporary technogenesis.* Chicago: University of Chicago Press.

Kuhn, T. S. 1962. *The structure of scientific revolutions.* Chicago: University of Chicago Press.

Kurke, L. 1999. *Coins, bodies, games, and gold: The politics of meaning in archaic Greece.* Princeton, NJ: Princeton University Press.

———. 2011. *Aesopic conversations: Popular tradition, cultural dialogue, and the invention of Greek prose.* Princeton, NJ: Princeton University Press.

Latour, Bruno. 1987. *Science in action: How to follow scientists and engineers through society.* Cambridge, MA: Harvard University Press.

———. 2003. *We have never been modern.* Cambridge, MA: Harvard University Press.

Mitchell, Timothy. 2011. *Carbon democracy: Political power in the age of oil.* New York: Verso.

Nicholson, Vivian, and Stephen Smith. 1977. *Spend, spend, spend.* London: Jonathan Cape.

Nielsen, M. A. 2012. *Reinventing discovery: The new era of networked science.* Princeton, NJ: Princeton University Press.

Nowotny, Helga, Peter Scott, and Michael Gibbon. 2011. *Rethinking science: Knowledge in an age of uncertainty.* Cambridge, UK: Polity.

Ratto, Matt. 2011. Critical making: Conceptual and material studies in technology and social life. *Information Society* 27(4): 252–260.

Ribes, David, and Thomas A. Finholt. 2009. The long now of technology infrastructure: Articulating tensions in development. *Journal of the Association for Information Systems* 10(5): 375–398.

Serres, Michel, ed. 1989. *Eléments d'histoire des sciences.* Paris: Bordas.

———. 1993. *Les origines de la géométrie: Tiers livre des* fondations. Paris: Flammarion.

Shinn, Terry. 1980. *Savoir scientifique et pouvoir social: L'École polytechnique—1794–1914.* Paris, Presses de la Fondation Nationale des Sciences Politiques.

Shirky, Clay. 2010. *Cognitive surplus: Creativity and generosity in a connected age.* New York: Penguin.

Siskin, Clifford, and William Warner. 2010. *This is enlightenment.* Chicago: University of Chicago Press.

Slota, Steven and Geoffrey C. Bowker. In prep. On the value of 'useless data': Infrastructures, biodiversity, and policy.

Snow, C. P. 1964. *The two cultures: And a second look.* Cambridge: Cambridge University Press.

Sohn-Rethel, Alfred. 1975. Science as alienated consciousness. *Radical Science Journal* 5: 65–101.

Sousanis, Nick. 2015. *Unflattening.* Cambridge, MA: Harvard University Press.

Star, Susan Leigh, and Karen Ruhleder. 1996. Steps toward an ecology of infrastructure: Design and access for large information spaces. *Information Systems Research* 7(1): 111–134.

Star, Susan Leigh. (1995) 2016. Misplaced concretism and concrete situations: Feminism, method and information technology. In *Boundary objects and beyond: Working with Leigh Star,* ed. Geoffrey C. Bowker, Stefan Timmermans, Adele Clarke, and Ellen Balka, 143–167. Cambridge, MA: MIT Press.

Stengers, Isabelle. 2011. *Thinking with Whitehead: A free and wild creation of concepts.* Cambridge, MA: Harvard University Press.

Tort, Patrick. 1999. *La raison classificatoire: Quinze études.* Paris: Aubier.

Tufte, E. R. 1983. *The visual display of quantitative information.* Cheshire, CT: Graphics Press.

9

Infrastructure, Potential Energy, Revolution

DOMINIC BOYER

Why This "Infrastructural Turn"?

Why is it that we are thinking and talking so much about infrastructure (the seminal texts are many but include Star and Ruhleder 1994, 1996; Bowker 2010; Anand 2012; Appel 2012; Gupta 2012; Harvey and Knox 2012, 2015; Barry 2013; Edwards et al. 2013; Larkin 2013; von Schnitzler 2013; Appel et al. 2015) in anthropology and related human sciences today? The first point to make is that infrastructure talk indexes a certain conceptual topology in which consequential epistemic or material substrata of experience are posited and mined for hidden truths (the Lévi-Straussian "structure," the Marxian "relations of production," are other examples). Anthropology, like other human sciences, claims that its expertise offers deep and extraordinary insight. Concepts like structure and now infrastructure operate generatively within our routine enterprise of revelation.

Yet, even if this first point is valid in a limited sense, it still does little to explain why we find a concept such as "infrastructure" so intuitive and compelling right now. So a second idea would be that infrastructure indexes the politics of the contemporary in some way. It is striking that the conceptual rise to intuitiveness of infrastructure roughly parallels the crisis and stasis of neoliberal governance since 2008. This hardly seems a coincidence, especially given that the Keynesianism that preceded neoliberalism, dominating western political economic theory and policy from roughly the mid-1930s until the mid-1970s, often utilized large-scale public works projects as key instruments for managing labor, "aggregate demand," and the affective ties of citizenship. Thirty years of privatization, financialization, and globalization later, this legacy of "public infrastructure" has become rather threadbare, capturing a general sense of evaporating futurity in the medium of corroded pipes and broken concrete. Of course, neoliberalism did promote aggressive investment and innovation in infrastructural systems necessary for the advance of financialization and globalization (not least telecommunications, the Internet, and transportation). At the same time, infrastructural temporalities look rather different from the perspective of the global South where, as Akhil Gupta (2013) has observed, ruination is a constant companion of infrastructure. But across the global North, one cannot be faulted for feeling a creeping sense of decay spreading across many infrastructural environments. Thus, the turn to infrastructure could be viewed as something like a conceptual New Deal for the human sciences—a return of the repressed concerns of public developmentalism to an academic environment that has, like much of the rest of the world, become saturated with market-centered messages and logics over the past three decades.

This is plausible too, but I fear the wishful thinking that either a conceptual or political return to the glory days of postwar modernity will offer us a way forward. Timothy Mitchell's *Carbon Democracy* (2011) argues convincingly that all the twentieth century's dominant models of economic ontology and health—classical, Keynesian, neoclassical, and neo-Keynesian—were equivalently indebted to apparatuses of carbon energy extraction and delivery, and of course to the magnitudes of power that could be coaxed from fossil fuels (see also Szeman, Wenzel, and Yaeger 2017). In particular, Mitchell argues that the Keynesian model of governance, with its characteristic belief that high levels of economic growth could be maintained through the medium of public investment and moderate welfarism, was only made possible by post–World War II imperial control over the Middle East's oil resources. This control guaranteed more or less free energy and power to the Anglo-American world and its allies, reigniting the fantasies of endless industrial growth that had echoed throughout the eras of coal, steam power, and atomic power before it. Yet, only a few decades later, with the formation of OPEC and the oil shocks of the 1970s, the geopolitics of carbon energy changed dramatically. More recently, "peak oil" projections—or, less controversially, the recognition that increasing fossil-fuel extraction is becoming dramatically more costly both economically and ecologically—have further disrupted the energic basis of growth-oriented economic theory and policy. It is thus not very surprising that since the 1970s, despite whatever neo-Keynesian or neoclassical policy package has been thrown at it, high levels of economic growth have not been restored to the West (Duménil and Lévy 2011). Mitchell argues that this model of "economic growth" and its attendant developmentalism was always premised upon oil as an infinite, inexpensive resource in the first place. The many subsequent wars waged or instigated in the Middle East by the Anglo-American powers and their allies have all desperately sought to recapture an oily infrastructure of growth, even as rising winds and waters, induced in large part by the burning of fossil fuels, begin to wreak serious havoc across the world.

This detour leads me to a third and final possible answer to the question of why infrastructure: Why now? I am coming to view the anthropology of infrastructure as an element of a much larger movement, what I would call, for lack of a less ugly term, the "anti-anthropocentric turn" in the human sciences (Boyer 2014). This turn followed a longish heyday of semiotic, psychological, and ideational trends in human-scientific theory, trends that synched closely with the halcyon years of Keynesian carbon modernity.

As carbon energopower entered into its first great crisis in the mid-1970s, the epistemic symptoms of paradigm shift were everywhere. Anarchist (punk) sentiment and experiment hissed through the cracks of Keynesian modernity. Marxian criticism also flourished for a time but had trouble reconciling itself with the fact that actually existing Marxist states promoted even greater social and ecological catastrophes. A paralyzed Left allowed neoliberalism to cement its ideological influence across the West. Second-wave and ecofeminism proved more difficult to silence, however, and these movements laid the foundation for decentering the intellectual and institutional legacies of postwar modernity. In the late 1970s, science and technology studies came of age and paid detailed critical attention to the epistemic legacies of the Enlightenment and practices of modern expertise. Figures such as Donna Haraway (1989) and Bruno Latour (1993) emerged as analysts of the contingencies of science, as prophets of the bankruptcy of modern nature/culture oppositions and as spokespersons for the "actancy" of animals, objects, and materials. The parallel rise of Foucauldian analysis of power/knowledge (e.g., 1979) further underscored a lost faith in modern technocratic imaginaries of science, growth, and development that had seemed so robust until the oil shocks changed everything.

Subsequently, posthumanism (e.g., Wolfe 2010) has challenged the human empire over other forms of life, the careless manipulation of companion species (Haraway 2003), and companion materials (Bennett 2010). And, in the past decade, we have seen a marvelous array of new conceptual movements working under banners such as "new materialism," "object-oriented ontology," "new realism," "speculative realism," and so on (see, e.g., de Landa 2002; Harman 2002; Meillassoux 2008; Morton 2010). Such thinking is far from homogeneous. But there is a family resemblance. Although philosophy typically believes itself undetermined by its socio-environmental context, a Hegelian might say that one cannot help but find the timing of these movements uncanny. They are all taking shape in the deepening shadow of the Anthropocene, an era in which one species—for the first time since cyanobacteria oxygenated the earth's atmosphere some 2.3 billion years ago—has proven itself capable of transforming the lifeworld of all species. That lifeworld, our lifeworld, is now failing. All the happy biopolitical promises, whether neoliberal or Keynesian, of endless growth, wealth, health, and productive control over "nature" now appear increasingly deluded and bankrupt, designs for Malthusian tragedy. Whatever more specific intellectual agendas today's anti-anthropocentric movements are pursuing, all of them index and oppose the toxic legacies of radically human-centered thinking and action. And thus, in their own way, I believe they offer commentaries on carbon modernity's accelerated death-bringing in the name of fueling a certain mode of human existence.

So, in one sense, our interest in infrastructure could be nothing more remarkable than the latest chapter in the good book of anthropological revelation. It may also signal a Keynesian nostalgia, summoning images from what is likely to have been the North's last great phase of imperial growth and prosperity. But, viewed in the context of the wider anti-anthropocentric turn in the human sciences, the conceptual promise of infrastructure seems to me a partial figuration of still deeper concerns about life in the Anthropocene. An optimistic view is that the turn toward infrastructure is a sign that we are conceptually re-arming ourselves for the struggle against the Anthropocene and the modernity that made it. The remainder of this chapter proceeds in this spirit of optimism.

Infrastructure as Potential Energy

Suppose we conceptualize infrastructure not in a nominal but rather in a modal way. That is, suppose we concentrate less on infrastructure as (often impressive, verging on sublime) material forms and more as an "energopolitical" (Boyer 2014) process, that which allows something to happen. Brian Larkin (2013) has laid rails for us here by describing infrastructures as "built networks that facilitate the flow of goods, people, or ideas and allow for their exchange over space" (328). He also observes that the "peculiar ontology" of infrastructures "lies in the facts that they are things and also the relation between things" (329).

Characterizing infrastructure in terms of mediation makes sense given not only what material infrastructures typically do (e.g., providing conduits for the transit of energy, bodies, and resources) but also in terms of the general conceptual emphasis on flow and friction in analytics of the post–cold war era of globalization (e.g., Appadurai 1996; Tsing 2005). Likewise, emphasizing that infrastructure is always a relation as well as a thing importantly underscores the fact that an infrastructure can never exist outside of a deictic (e.g., *infra* "below") relationship to something else. But let us push this insight a bit far-ther. What Larkin is really saying is that whatever else an infrastructure might be it must always serve as the foundation that enables something else to happen (and, as Larkin rightly points out, be enabled to enable in this way). Infrastruc-ture then becomes "political" in the specific modal sense conjured by Latinate terms like *pouvoir* and *poder*. What do infrastructures do? They enable. And what allows them to enable? The storage and expenditure of energy. Infrastruc-tures can thus be viewed in terms of "potential energy."

To better explain what I mean by this, I will begin with an enabling rela-tionship that is already well known: labor and capital in the classic Marxian formulation. Capital for Marx was a dimension of the objectification of human labor power, specifically, a result of how the division of labor severed labor's capacity to channel human will in the development of the self. Instead of an ideal dialectical process of self-realization, capital congealed labor power in a way that could circulate beyond the self, be appropriated and commanded by others, and thus be transformed into new social and material forms. Capital was in this way a means of remote enablement (yet one always enabled *de infra* by labor power). And, once it was set into motion on a mass scale and stabi-lized by institutions such as money and wage labor, quantifiable appropriated labor-time became the logic of social value in modern society. As Marx wrote in the first chapter of *Das Kapital*, "As values, all commodities are only definite masses of congealed labor-time."

Here there is an interesting twist to the story we all know. "Congelation," from the Latin verb *congelare*, "to freeze together," is actually a slightly misleading translation of the actual noun Marx uses, *Gallert*, which refers to a gelatinization process in which different animal substances with the potential to yield glue (e.g., meat, bone, connective tissue) are boiled and then cooled to produce a "semisolid, tremulous mass . . . a concentrated glue solution" (Sutherland 2008). Rather than a freezing together of independent parts, *Gallert* indicates an ontological transformation through adding and then subtracting thermal en-ergy, a recipe of different fleshy forms rendered through heating and cooling into a single sticky material: human labor-in-the-abstract binding commodi-ties, people, machines, and "nature" together with its glue. Indeed, one might go a step farther and say that this glue potential was unlocked by the thermal rendering process itself.

Paul Burkett and John Foster (2006) argue that this is more than just meta-phor. Their view is that there is a powerful energo-metabolic substrate to Marx's theories of labor power, alienation, and value extraction. On the topic of surplus value, they write,

Of course, this value (energy) surplus is not really created out of nothing. Rather, it represents capitalism's appropriation of portions of the *potential* work embodied in labor power recouped from metabolic regeneration largely during non-worktime. And this is only possible insofar as the regeneration of labor power, in both energy and biochemical terms, involves not just consumption of calories from the commodities purchased with the wage, but also fresh air, solar heat, sleep, relaxation, and various domestic activities necessary for the hygiene, feeding, clothing, and housing of the worker. Insofar as capitalism forces the worker to labor beyond necessary labor time, it encroaches on the time required for all these regenerative activities. (127)

In this context, capital becomes an appropriation quite literally of fleshy power, a sapping and storage of the regenerative potential of being. The authors go on to conclude more generally:

> What Marx and Engels generated in their historical-dialectical materialism was a theory of the capitalist labor, production, and accumulation process that was not only consistent with the main conclusions of thermodynamics originating in their time, but also extraordinarily amenable to ecological laws. . . . At the same time Marx developed a sophisticated theory of the metabolic character of the human labor process and of the metabolic rift that appears within capitalism. This analysis not only recognized that "matter matters" but was sensitive to the biochemical processes of life itself and to emerging evolutionary theory. (144)

Porting this analytic strategy over to the *Grundrisse* and the second volume of *Kapital*, where Marx describes an ineluctable evolution of circulating capital toward fixed capital and of fixed capital toward automated machinery, we come closer to a portrait of infrastructure as potential energy-storage system, as a means for gathering and holding productive powers in technological suspension.

In Marx's vision, capital strives across its historical development to make itself independent of labor, to be able to absorb the productive powers of labor into itself. As one might expect of the logic of bourgeois economy, capital seeks its liberty. The development of fixed capital (the part of the production process that retains its use-form over a period of time rather than being wholly consumed in a production process—that is, a machine, or even cattle pulling a plow, but not the coal burned to produce steam) is the first stage of this process. Marx emphasizes that durability is crucial: fixed capital must *durably* stand in for direct human labor. But the more decisive phase is the movement from fixed capital toward automated machinery, a productive apparatus that operates mechanically according to human design and in which "the human being comes to relate more as watchman and regulator to the production process... instead of being its chief actor" (Marx [1861] 1974: Notebook VII).

Automation not only advances capital's desire to durably emancipate itself from labor but also precipitates the final paradox between exchange value and use value that Marx believed would necessitate the eventual collapse of the capitalist mode of production:

> On the one side, then, [capital] calls to life all the powers of science and of nature, as of social combination and of social intercourse, in order to make the creation of wealth independent (relatively) of the labor time employed on it. On the other side, it wants to use labor time as the measuring rod for the giant social forces thereby created, and to confine them within the limits required to maintain the already created value as value. Forces of production and social relations—two different sides of the development of the social individual—appear to capital as mere means, and are merely means for it to produce on its limited foundation. *In fact, however, they are the material conditions to blow this foundation sky-high.* ([1861] 1974: Notebook VII, emphasis added)

In Marx's model, what we would call "infrastructure" begins as the outer skin of capital like the steel casing of a pipeline that allows precious liquid to flow through its center. But over time, the labor power invested into the technological development of automated machinery hardens into a solid form that is both persistent and nearly autonomously productive of useful things. This apparatus is not, as it were, a *perpetuum mobile*, generating productivity from nothing. Its productivity depends wholly on the gelatinized combination of expertise, activity, materials, and forces that have filled the mold of its infrastructural design. In other words, infrastructure stores the productive powers of labor—mental, material, natural, to use Marx's categories—in such a way that they can be released later in magnitudes that appear to transcend nominal inputs. Technology, as productive infrastructure, thus appears to be capable of generating and distributing use values with limited need for direct (human) labor power. This imaginary persists into the twenty-first century and not only among critics of capital. There is, for example, a TransCanada brochure for its gleaming white Keystone XL pipeline (see figure 9.1) that proudly announces how its 41 pumping stations, designed to move 830,000 barrels of crude oil a day across Canada and the United States, "will be remotely monitored and

Keystone XL Pipeline
Pump Stations

FIGURE 9.1 A Keystone XL Pipeline pumping station brochure, TransCanada.

controlled, 24 hours a day, seven days a week, from the Oil Control Center (OCC) in Calgary" and visited only occasionally by technicians and able to "safely shut down" by itself. The one technician in the image is pressed to the margins, back turned toward the camera, observing a wall of screens. Here is infrastructure, containing "all the powers of science" and engineering, able to enable so many modern projects, going it (almost) solo.

Even without an extensive Marxian detour, I believe that it makes intuitive sense to view infrastructure as "stored" or potential energy in that it is no secret that a great variety and density of distinct forms of activity go into the production and maintenance of the sort of mediation systems we normally conceive as infrastructure (e.g., roads, pipelines, grids), designed to facilitate flows and to enable other things to happen. We need not and indeed should not be limited by a Marxian focus on commodity production in this respect. Neither a pipeline nor an electrical grid directly produces a commodity, but without their mediational apparatuses, what flows through them could scarcely be commoditized, let alone could their energic vitalities be brought to the making of so many other useful things. Pipelines and grids are thus also *Gallerte* in their own right. The ingredients of a medium- or high-voltage grid, for example, include not only expertise in electrical engineering, and expert construction labor, but also the legacies of centuries of materials science and manufacture (Hughes 1983) that have provided serviceable steel, concrete, aluminum, copper, reinforced plastics, ceramics, tempered glass, silicone rubber, carbon, and glass fiber for the making of pylons, power cables, transformers, and insulation. In the end, after considerable rendering, these elements look to us, especially at a distance, as nothing more remarkable than a metal fabric of vertical and horizontal lines—the familiar mesh of grid, with what Cymene Howe terms a "gridlife" (Howe n.d.) all its own.

But the detour is also not without purpose. The reason I think a Marxian path to rethinking infrastructure is critically relevant to the Anthropocene is its allowal for the possibility that infrastructure can not only enable a certain arrangement of other things to happen but that its potential energy can also blow the very same arrangement "sky-high." Infrastructure, in other words, can be revolutionary as well as conservative, a lesson not lost on Lenin, who famously proclaimed to the Moscow conference of the Russian Communist Party in 1920 that "Communism is Soviet power plus the electrification of the whole country."

As we come to understand our position in the Anthropocene, we desperately need to consider and where possible to enable this revolutionary potential of infrastructure. At the speed that climate and earth science tell us action is required to evade the worst scenarios of global warming, an incremental, partial, slow transition away from fossil-fuel sources and infrastructures is simply not a luxury we can afford. We thus need to discover what I have termed "revolutionary infrastructure" (Boyer 2017). But we also need to consider that whatever revolutionary infrastructure may be, it may not resemble the familiar infrastructures of carbon modernity that currently encompass and enable us.

Infrastructure and Revolution

It takes just a thirty-minute boat ride down what is perhaps the densest corridor of carbon energy infrastructure in the world, the Houston Ship Channel, to witness the massive supertankers loading liquid natural gas for sale abroad, the petroleum refineries the size of small towns, the sprawling plastics industries, even the spectacular hills of recycled materials and wind turbines from Denmark being unloaded to be shipped to West Texas, to gain an appreciation for the colossal energopolitical apparatus that would need to be unmade or remade to make a serious effort at addressing the Anthropocene. It is an awesome and dispiriting journey, not a little sublime. Figure 9.2 is an image I like to call "BodyWorld United States," its plastinated blue veins and red arteries (natural gas and oil pipelines) made visible, all leading back to the beating heart of carbon energopower in Louisiana, Oklahoma, and Texas. Never has the coastal

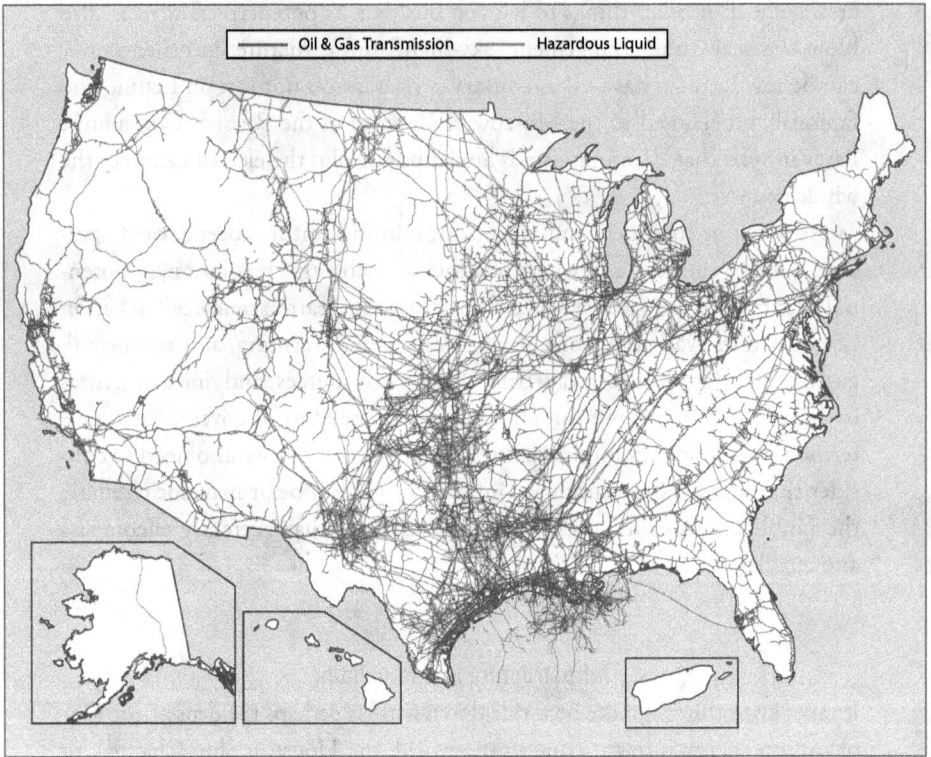

FIGURE 9.2 Map of U.S. pipelines. Data courtesy Pipeline and Hazardous Materials Safety Administration.

United States—with all its supposed loci of financial, cultural, and governmental power—seemed more peripheral to national political infrastructure.

Climate and earth scientists are issuing warnings of increasing urgency and unanimity that what amounts to revolutionary action is needed to prevent the acceleration of global warming and the ecological catastrophes that warming is predicted to entail. The scientists do not always use the language of revolution but they might as well. And some, like Kevin Anderson, the former director of the Tyndall Centre for Climate Change Research at Manchester University, are willing to speak bluntly: "After two decades of bluff and lies, the remaining 2°C budget demands revolutionary change to the political and economic hegemony."[1]

We human scientists understand—even if we perhaps underestimate the deep phenomenological roots of the problem—the extent to which this revolution is impeded by the convenience and luxury of an energy-intensive modernity that has been enabled by fossil fuels since its beginning. But we are also less obviously hampered by the fact that our political models of revolution (at least since the mid-nineteenth century) have been closely aligned with the harnessing of still more massive magnitudes of energy and industrial productivity in order to catalyze a breakthrough in the fabric of the bourgeois capitalist order. It may well be the case that Marx's critique of capital was more sensitive to energo-metabolic questions than we had previously thought but his vision for postcapitalist society contains no consideration of the possibility of human action permanently changing or damaging natural order, let alone rumination upon specific anthropocenic processes such as greenhouse gas emission, ocean acidification, and species extinction. Not unlike the bourgeois political economists he otherwise roundly condemned, Marx appears to have thought that industrialization was an intrinsically historically progressive development in its creation of a technological basis for the automated production of commodities, an abundance of useful things that could unite and benefit the entire (human) species after the negation of capital's pernicious effort to liberate itself at the expense of the planet (an effort captured well in Malm's concept of "capitalocene" and the conversations that have trailed it; e.g., Malm 2013; Malm and Hornborg 2014; Haraway 2015; Moore 2015). Ironically, from the perspective of contemporary eco-Marxism, Marx himself stands more on the side of contemporary apologists for fossil fuels as an optimist regarding modern infrastructure's ability to produce technological solutions to the problems generated by the metabolics of modern industry.

It is unsurprising therefore that massive energo-infrastructural growth was central to the Leninist-Bolshevist revolutionary state as well. Indeed, the

1920s rural electrification campaign served as a prototype for the Stalinist five-year plan of organized economy. It is worth hearing a little more from Lenin's speech to the Moscow conference of the Russian Communist Party to understand why electrification was such a priority:

> Communism is Soviet power plus the electrification of the whole country, since industry cannot be developed without electrification. This is a long-term task which will take at least ten years to accomplish, provided a great number of technical experts are drawn into the work. A number of printed documents in which this project has been worked out in detail by technical experts will be presented to the Congress. We cannot achieve the main objects of this plan—create so large regions of electric power stations which would enable us to modernize our industry—in less than ten years. Without this reconstruction of all industry on lines of large-scale machine production, socialist construction will obviously remain only a set of decrees, a political link between the working class and the peasantry, and a means of saving the peasants from the rule by Kolchak and Denikin; it will remain an example to all powers of the world, but it will not have its own basis. Communism implies Soviet power as a political organ, enabling the mass of the oppressed to run all state affairs—without that, communism is unthinkable. We see proof of this throughout the world, because the idea of Soviet power and its program are undoubtedly becoming victorious throughout the world. We see this in every phase of the struggle against the Second International, which is living on support from the police, the church and the old bourgeois functionaries in the working-class movement.
>
> This guarantees political success. Economic success, however, can be assured only when the Russian proletarian state effectively controls a huge industrial machine built on up-to-day technology; that means electrification. For this, we must know the basic conditions of the application of electricity, and accordingly understand both industry and agriculture. (Lenin 1965 [1920])

Without electrification enabling the constitution of a massive industrial apparatus, communism is "unthinkable." This, of course, does not stop with communism. The making of a huge electrified "industrial machine" was the enormous multi-ideological project of the twentieth century that has now spilled over into the twenty-first century, always expanding on a global basis under the banner of "development" (Boyer 2015; Gupta 2015). All the dominant economic and political philosophies of the past century have made their contributions to expanding and intensifying this apparatus: liberalism, social-ism, Keynesianism, neoliberalism. None, interestingly, seems to have doubted that an expanded industrial apparatus would give their philosophy political-economic leverage over the others. It may be the sole issue they all agreed upon.

So we return to the problem of an impasse in what might be called "epis-temic infrastructure." Familiar paradigms of revolutionary action offer us little hope for remediating anthropocentric conditions swiftly and broadly. I would thus like to spotlight a less familiar and perhaps in some respects less dramatic model of revolution: Hermann Scheer's call for a decentralized "solar econ-omy" (2004). Scheer was an interesting figure, for thirty years a parliamentar-ian in Germany's Social Democratic Party, a revolutionary thinker in a party that had long lost its taste for radical politics. He was one of the chief politi-cal architects of Germany's *Energiewende* (renewable energy transition) and cowrote Germany's famous feed-in tariff law that forced utilities to guarantee long-term above-market-price electricity purchase agreements for renewable energy to create stability and incentives for small solar and wind power pro-ducers. In 2015, 31 percent of gross electricity production in Germany came from renewable resources, in large part thanks to the momentum established by Scheer's feed-in tariffs. In the past fifteen years, versions of the German legis-lation have been adopted by many other countries. The feed-in tariff has proved to be perhaps the single most effective policy instrument for catalyzing rapid investment in renewable energy production despite facing concerted resistance from electricity utilities and the fossil-fuel industry.

Scheer himself ruminated at great length about the reasons for resistance. He argued it was more than simply a desire to maintain a profitable indus-try. Scheer pointed to the infrastructural adaptation of global and national economies to the "long supply chains" characteristic of fossil- and nuclear-fuel resources. Anticipating Mitchell's more recent analysis of "carbon democracy" (2011), Scheer viewed twentieth- and twenty-first-century globalization as largely driven by the extraction and control of these fuels. Scheer observed that long energy supply chains are, in their very nature, inefficient and thus demand infrastructures of translocal domination to guarantee unimpeded flows of critical resources. This domination imperative has been masked by nation-alist discourses of security and well-being, by post/neo/colonial missions of civilization and development, and most recently by the utopian logic of a self-regulating "market." Scheer wrote, "Being invisible, the hand of the market can steal and exploit without being recognized. The result is not harmony, but tension, division and disruption. My proposal, to rely above all on the *visible hand of the sun* . . . is more precise, more comprehensive . . . more appropriate to people's needs and more realistic. It is also free of danger and definitely less utopian" (2004: 33).

Solar energy, whether in its direct form of insolation or indirectly in the forms of wind and biomass, has the physical advantages, Scheer argued, of ubiquity and superabundance, thus allowing for more efficient and decentralized short supply chains that are also more susceptible to democratic political control: "Shorter renewable energy supply chains will make it impossible to dominate entire economies. Renewable energy will liberate society from fossil fuel dependency" (2004: 89).

Recognizing that reliance upon fossil and nuclear fuels is driving the world to anthropocentric ruin, Scheer challenges the notion that we need to maintain large-scale power grids and pipeline systems at all. He calculates that even energy-intensive modernity can be maintained purely on the basis of small-scale solar resources given contemporary technologies. The problem is that grids and pipeline systems—products of early twentieth-century political and industrial concentration enabled in turn by the burning of fossil fuels—have become a chief instrument in the monopolization of political authority, an "energopolitical" (my term, not Scheer's) apparatus mutually reinforcing the inertia of a particular organization of fuel and a particular organization of political power. Their energopolitics constitutes an energo-material path dependency while also casting a dark shadow of improbability over any imagined alternative to the long-chained fossil status quo. Carbon epistemics thus flourish, reinforcing the naturalness and inevitability of carbon infrastructure. Grids and pipelines are allowed to hide their inefficiencies through inflated "black-boxed" pricing and unwarranted public subsidies, which for Scheer in essence amount to vassalage fees paid by the population to the masters of fossil fuels and centralized authority. Although Scheer is sympathetic to Marxian revolution in certain respects, he argues that a solar energy revolution will act to dissolve centralized into distributed systems: "Comprehensive use of renewable energy would take the wind from the sails of an economic globalization and industrial concentration process driven by the scarcity of fossil-fuel reserves. This alone would spark a process of de-concentration, de-monopolization and the re-regionalization of economic structures" (2004: 86). He observes that the imagined freedoms of internetworked society have never come to pass precisely because the rise of the Internet and social mediation does not challenge energopolitical concentration per se.

Scheer offers other intriguing policy proposals including the destabilization of translocal high-voltage grids through the creation of public municipal energy corporations (2004: 269–273) that would control highly localized energy production and storage systems and be more responsive to local interests and needs. "For society [this] means no long-distance power cables, apart from the few lines needed to bring power from large dams and windfarms. As the centralized gas and electricity industry loses its purpose, and municipal power supplies and energy self-sufficiency expand into more and more domains, the high- and medium-voltage cables will gradually disappear. Pylons will no longer march across the landscape, more than compensating for any decline in landscape quality due to wind turbines" (2004: 273).

The broad strokes of Scheer's economic philosophy are neo-physiocratic, an argument for reestablishing the "primary economy" of agriculture and forestry over industry and services as the center of economic life. But, like Marx and Lenin, he seeks not a retreat backward into pre-industrial conditions but rather a revolutionary leap forward into a new form of sociality, one that is energy intensive and technologically enabled but resolutely local, sustainable, and diverse.

In this respect Scheer's model of solar economy resonates strongly with a variety of other recent social movements organized in opposition to climate change and large-scale society and industry, for example the "degrowth" movement, the "cooperative economy" movement, and perhaps especially "transition culture."[2] These movements pass beyond policy arguments and toward what one could describe as anti-anthropocenic pastoralism. Speaking with members of the Transition Houston movement, for example, I have been struck by their commitment to work not just on sustainability initiatives but on the affects induced by "coming to terms with the reality of climate change." The Transition US website states, "The challenges we face are not just caused by a mistake in our technologies but are a direct result of our world view and belief system. The impact of the information about the state of our planet can generate fear and grief—which may underlie the state of denial that many people are caught in. Psychological models can help us understand what is really happening and avoid unconscious processes sabotaging change (e.g., addictions models, models for behavioral change). This principle also honors the fact that Transition thrives because it enables and supports people to do what they are passionate about, what they feel called to do."

Likewise, the Scheerian cathexis of local political authority and accountability echoes with my recent research experiences in Reykjavík and the Isthmus of Tehuantepec in which neo-anarchist movements have gained substantial strength, not directly in opposition to carbon energy it is true, but certainly in opposition to the forms of centralized political and economic infrastructure that Scheer argues are so deeply entwined with carbon energy. Members of Iceland's Best Party directly linked their ability to experiment with new direct democratic practices to the city's abundant and relatively cheap geothermal energy supply. Meanwhile in southern Mexico, popular assemblies have risen in indigenous ikojts and binnizá communities to resist the densest concentration of on-shore wind-power development anywhere in the world. Their argument is not so much that wind power does not represent a preferable alternative to conventional fuel supplies at a global level. Rather, they argue that it is not their responsibility to sacrifice their land and wind to an organization of "clean energy" development that is driven not by local needs and interests but rather by international capital seeking high rates of return, a national government's clean electricity targets, by transnational corporations seeking to green their image, and a seemingly unquestioned desire to grow a grid-based energy supply for commercial and industrial purposes. "All this supposed clean energy," one activist told us, "is going to power more Walmarts and cement factories, and those are the true problem."

This spirit has also animated the Center for Energy and Environmental Research in the Human Sciences (CENHS) collective at Rice University, which was organized in 2013 to bring the arts, humanities, and social sciences into conversation and action regarding our contemporary energy and environmental challenges.[3] CENHS is a self-conscious effort to generate new epistemic infrastructure, some of it revolutionary. The collaborative research process gives us a more precise critical-analytical grasp on the impasses that enroll us and also room for new creative experiments, speculative imaginings, and affective attachments. Above all, the collective seeks to offer a beacon, a rallying point, and a safe space in which to find the courage and determination to act when the scale of the problem seems utterly negating. The anarcho-surrealists of Iceland's Best Party used to describe their organization less as a political party and more as a self-help movement, "like Alcoholics Anonymous. We try to take one day at a time, to not overreach our boundaries and to maintain joy, humility and positive thinking. . . . We often say that we aren't doing what we want to do but what needs to be done." These are the operational affects of the CENHS collective, too: less earnest moralizing (we hope) and more laughter, perseverance, and humility in the process of doing things that need to be done. And where better to do those things than in Houston—possibly the least sustainable city on the planet, the beating heart of twentieth-century petroculture, the throne of fossil necrocracy?[4] We are the hobbits inching up its Mount Doom.

Conclusion: Toward Revolutionary Infrastructure

The promise of revolutionary infrastructure is to hack and redistribute the stored energies of anthropocentric carbon modernity toward projects of imagining and pursuing paths out of the Anthropocene. Some of these paths may speak a familiar language of verdant becomings, of "smart green cities" and the like (e.g., Clark and Cooke 2016). I think we may be grateful for their design work and pastoral efforts to reassure an increasingly anxious humanity that a rapid transition toward carbon neutrality lies safely within our grasp. However, they may underestimate the extent to which we humans deeply desire a transition in which (a) technology does most of the work for us and (b) we, especially the global North, get to hold on to the abundance we accumulated during the fossil era. This desire seems to me hallucinatory at a fundamental level; sacrifice, transformational work, and downscaling the size and impact of humanity obviously await us.

So I will make a case for more dissonant paths toward a process of personal and civilizational rebecoming. Riffing on his concept of "hyperobjects"— objects too massive and multiphasic in their distribution in time and space for humans to fully comprehend or experience them in a unitary way, like global warming and capitalism and plastic bags—Timothy Morton and I have recently written (2016) of the need to confront and reform the transcendence-seeking "hypersubjects" (usually but not exclusively white, straight, northern males) that gifted the world the Anthropocene as part of their centuries-long project of remaking the planet for their own convenience and luxury. As hypersubjects ourselves, we understand intuitively the circuits of desire that ramify the condition that Sara Ahmed has brilliantly diagnosed as "white men" and the truth she has spoken, that "it takes conscious willed and willful effort not to reproduce an inheritance." Because fossil fuels have long enabled the dominion of hypersubjects, part of this inheritance is the reproductive infrastructure of carbon modernity, which, really, in our view, ought to be called something like "androleukoheteropetromodernity," to vomit out an appropriately ugly word for an ugly era.

"White men" need reform. We have never listened to the nonwhite men telling us that, although that is precisely what they have told us for as long as there have been white men. Now the titans we have created—the hyperobjects— have shattered our distinctive fantasies of command and control. The hyperobjects loom and leer and cast deepening shadows. At first we deny they are there. Then we say they are not so dangerous after all. But we are hearing their voices in our head now whispering that our time is ending. Here is a dark secret:

we are prepared to take the whole planet down with us because it is far easier for us to imagine death than it is to imagine not receiving the due share of our inheritance, perpetuating the dominion of our kind. So for the moment we lash out, we pout and vilify, we bargain for salvational machines and afterlife redemptions.

But here is another truth: the time of hypersubjects is ending and the time of hyposubjects is beginning. I cannot say whether "white men" will eventually prove themselves capable of reform but if they do not they will find themselves increasingly irrelevant to what comes next. Revolutionary infrastructure is the process of unmaking androleukoheteropetromodernity, of squatting in its ruins, of practicing subscendence rather than transcendence. Revolutionary infrastructure will not register on the androleukoheteropetromodern radar at first; that is to say, to contemporary optics and epistemics, the idea of revolutionary infrastructure seems preposterous, utopian, and fantastic. Over time though, through the work of proliferating decentralized small-scale action, the carbon gridworld will find itself incrementally disabled by the tapping and redistribution of its materials and energies. This is not to say this will be a peaceful transition of powers or that it will not be accompanied by turbulence, even "heat death" (Parisi and Terranova 2000). But evidence of the rebecoming of human life beyond grids and fuels is there for those who seek it; designs and prototypes for revolutionary infrastructure already abound, Scheerian and otherwise, seeking recognition and traction, aspiring to be durably enabled to enable the ending of the Anthropocene. Please do what you can to help.

NOTES

1 See http://kevinanderson.info/blog/why-carbon-prices-cant-deliver-the-2c-target.
2 See, e.g., the Alternative Festival for Solidarity and Cooperative Economy (https://www.facebook.com/festival4sce), the Biannual International Conference on Degrowth since 2008, and Transition Network (http://www.transitionnetwork.org).
3 See the CENHS website (http://culturesofenergy.org).
4 See *Necrocracy* (http://www.o-matic.com/play/necrocracy).

REFERENCES

Ahmed, Sara. 2014. White men. *feministkilljoys* blog. https://feministkilljoys.com/2014/11/04/white-men.

Anand, Nikhil. 2012. Pressure: The PoliTechnics of water supply in Mumbai. *Cultural Anthropology* 26(4): 542–564.

Appadurai, Arjun. 1996. *Modernity at large: Cultural dimensions of globalization*. Minneapolis: University of Minnesota Press.

Appel, Hannah. 2012. Offshore work: Oil, modularity, and the how of capitalism in Equatorial Guinea. *American Ethnologist* 39(4): 692–709.

Appel, Hannah, Nikhil Anand, and Akhil Gupta. 2015. Introduction: The infrastructure toolbox. *Cultural Anthropology* website. https://culanth.org/fieldsights/714 -introduction-the-infrastructure-toolbox.

Barry, Andrew. 2013. *Material politics: Disputes along the pipeline*. New York: Wiley.

Bennett, Jane. 2010. *Vibrant matter: A political ecology of things*. Durham: Duke University Press.

Boyer, Dominic. 2014. Energopower: An introduction. *Anthropological Quarterly* 87(2): 309–333.

———. 2015. Anthropology electric. *Cultural Anthropology* 30(4): 531–539.

———. 2017. Revolutionary infrastructure. In *Infrastructures and social complexity: A companion*, ed. Penelope Harvey, Casper Bruun Jensen, and Atsuro Morita, 174–186. New York: Routledge.

Boyer, Dominic, and Timothy Morton. 2016. Hyposubjects. *Cultural Anthropology* website. https://culanth.org/fieldsights/798-hyposubjects.

Bowker, Geoffrey C. 2010. Towards information infrastructure studies: Ways of knowing in a networked environment. In *The international handbook of internet research*, ed. Jeremy Hunsinger, Lisbeth Klastrup, and Matthew Allen, 97–117. New York: Springer.

Burkett, Paul, and John Bellamy Foster. 2006. Metabolism, energy and entropy in Marx's critique of political economy: Beyond the Podolinsky myth. *Theory and Society* 35: 109–156.

Clark, Woodrow, II, and Grant Cooke. 2016. *Smart green cities: Toward a carbon neutral world*. New York: Routledge.

de Landa, Manuel. 2002. *Intensive science and virtual philosophy*. New York: Continuum.

Duménil, Gérard, and Dominique Lévy. 2011. *The crisis of neoliberalism*. Cambridge, MA: Harvard University Press.

Edwards, Paul N., et al. 2013. *Knowledge infrastructures: Intellectual frameworks and research challenges*. Ann Arbor, MI: Deep Blue.

Groeger, Lena. 2012. Pipelines explained: How safe are America's 2.5 million miles of pipelines? *ProPublica.org*. http://www.propublica.org/article/pipelines-explained-how-safe -are-americas-2.5-million-miles-of-pipelines.

Gupta, Akhil. 2012. *Red tape: Bureaucracy, structural violence, and poverty in India*. Durham: Duke University Press.

———. 2013. Ruins of the future. Paper delivered at the American Anthropological Association Annual Meeting, Chicago, IL, November.

———. 2015. An anthropology of electricity from the global South. *Cultural Anthropology* 30(4): 555–568.

Haraway, Donna. 1989. *Primate visions: Gender, race, and nature in the world of modern science*. New York: Routledge.

———. 2003. *The companion species manifesto: Dogs, people and significant otherness*. Chicago: Prickly Paradigm.

———. 2015. Anthropocene, capitalocene, plantationocene, chthulucene: Making kin. *Environmental Humanities* 6: 159–165.

Harman, Graham. 2002. *Tool-being: Heidegger and the metaphysics of objects.* Chicago: Open Court.

Harvey, Penny, and Hannah Knox. 2012. The enchantments of infrastructure. *Mobilities* 7(4): 521–536.

———. 2015. *Roads: An anthropology of infrastructure.* Ithaca, NY: Cornell University Press.

Howe, Cymene. N.d. Gridlife. Unpublished manuscript, Department of Anthropology, Rice University.

Hughes, Thomas P. 1983. *Networks of power: Electrification in Western society, 1880–1930.* Baltimore, MD: Johns Hopkins University Press.

Larkin, Brian. 2013. The politics and poetics of infrastructure. *Annual Review of Anthropology* 42: 327–343.

Latour, Bruno. 1993. *We have never been modern.* Cambridge, MA: Harvard University Press.

Lenin, Vladimir Ilich. 1965 [1920]. Our foreign and domestic position and party tasks. In *Lenin's collected works,* vol. 31, 4th English ed., 408–426. Moscow: Progress. https://www.marxists.org/archive/lenin/works/1920/nov/21.htm.

Malm, Andreas. 2013. The origins of fossil capital: From water to steam in the British cotton industry. *Historical Materialism* 21(1): 15–68.

Malm, Andreas, and Alf Hornborg. 2014. The geology of mankind? A critique of the Anthropocene narrative. *Anthropocene Review* 1(1): 62–69.

Marx, Karl. (1861) 1974. *Grundrisse der Kritik der Politischen Ökonomie.* Berlin: Dietz Verlag.

Meillassoux, Quentin. 2008. *After finitude: An essay on the necessity of contingency.* Trans. Ray Brassier. London: Continuum.

Mitchell, Timothy. 2011. *Carbon democracy: Political power in the age of oil.* New York: Verso.

Moore, Jason. 2015. *Capitalism in the web of life.* New York: Verso.

Morton, Timothy. 2010. *The ecological thought.* Cambridge, MA: Harvard University Press.

———. 2013. *Hyperobjects: Philosophy and ecology after the end of the world.* Minneapolis: University of Minnesota Press.

Parisi, Luciana, and Tiziana Terranova. 2000. Heat-death: Emergence and control in genetic engineering and artificial life. In *CTHEORY,* ed. Arthur Kroker and Marilouise Kroker. http://www.ctheory.net/articles.aspx?id=127

Scheer, Hermann. 2004. *The solar economy: Renewable energy for a sustainable global future.* London: Earthscan.

Schwenkel, Christina. 2013. Post/socialist affect: Ruination and reconstruction of the nation in urban Vietnam. *Cultural Anthropology* 28(2): 252–277.

Star, Susan Leigh. 1999. The ethnography of infrastructure. *American Behavioral Scientist* 43(3): 377–391.

Star, Susan Leigh, and Karen Ruhleder. 1994. Steps towards an ecology of infrastructure: Complex problems in design and access for large-scale collaborative systems. *Proceed-*

ings of the Computer Supported Cooperative Work Conference, Chapel Hill, NC, October 22–26. 253–264.

———. 1996. Steps toward an ecology of infrastructure: Borderlands of design and access for large information spaces. *Information Systems Research* 7(1): 111–134.

Sutherland, Keston. 2008. Marx in jargon. *World Picture*. http://worldpicturejournal .com/WP_1.1/KSutherland.html.

Szeman, Imre, Jennifer Wenzel, and Patricia Yaeger, eds. 2017. *Fueling culture: 101 words for energy and environment*. New York: Fordham University Press.

Tsing, Anna. 2005. *Friction: An ethnography of global connection*. Princeton, NJ: Princeton University Press.

von Schnitzler, Antina. 2013. Traveling technologies: Infrastructure, ethical regimes, and the materiality of politics in South Africa. *Cultural Anthropology* 28(4): 670–693.

Wolfe, Cary. 2010. *What is posthumanism?* Minneapolis: University of Minnesota Press.

CONTRIBUTORS

NIKHIL ANAND is Assistant Professor of Anthropology at the University of Pennsylvania. His research explores the political ecology of urban infrastructures and the social and material relations that they entail. His first monograph, *Hydraulic City: Water and the Infrastructures of Citizenship in Mumbai* (Duke University Press, 2017), explores the quotidian way in which cities and citizens are made through the management of water pipes. He is a member of the Faculty Working Group in Environmental Humanities at the University of Pennsylvania. His new work examines the life of seas in times of climate change.

HANNAH APPEL is Assistant Professor of Anthropology at the University of California, Los Angeles. Her research explores the daily life of capitalism—from the private sector in Africa; to the relationships among finance, debt, and activism in the United States; to the reemergent dialogue between economics and anthropology. She is a coeditor (with Arthur Mason and Michael Watts) of *Subterranean Estates: Lifeworlds of Oil and Gas*, and she is finishing an ethnography entitled *Oil and the Licit Life of Capitalism in Equatorial Guinea*, which explores the U.S.-based oil and gas industry's efforts to disentangle the production of profit from the frictions of place. The manuscript dwells on questions of infrastructure, the contract and the corporate form, and the ethnographic life of Equatorial Guinea's national economy.

GEOFFREY C. BOWKER is Professor at the School of Information and Computer Science, University of California at Irvine, where he directs the Evoke Laboratory, which explores new forms of knowledge expression. His recent positions include Professor of and Senior Scholar in Cyberscholarship at the University of Pittsburgh iSchool, and Executive Director, Center for Science, Technology and Society, Santa Clara. Together with Susan Leigh Star he wrote *Sorting Things Out: Classification and Its Consequences*; his most recent books are *Memory Practices in the Sciences* and the edited collection *Boundary Objects and Beyond: Working with Leigh Star* (with Stefan Timmermans, Adele Clarke, and Ellen Balka). He is currently working on big data policy and on scientific cyberinfrastructure, as well as completing a book on social readings of data and databases. He is a founding member of the Council for Big Data, Ethics and Society.

DOMINIC BOYER is Professor of Anthropology at Rice University and Founding Director of the Center for Energy and Environmental Research in the Human Sciences (culturesofenergy.org), the first research center in the world designed specifically to promote research on the energy/environment nexus in the arts, humanities, and social sciences. He is part of the editorial collective of the journal *Cultural Anthropology* (2015–2018), and he also edits the Expertise: Cultures and Technologies of Knowledge book series for Cornell University Press. His most recent monograph is *The Life Informatic: Newsmaking in the Digital Era*. With James Faubion and George Marcus, he recently edited *Theory Can Be More Than It Used to Be* and with Imre Szeman he developed *Energy Humanities: An Anthology*. His next book, *Energopolitics*, is part of a collaborative multimedia duograph with Cymene Howe, which will explore the complexities of wind-power development in southern Mexico. With Howe, he also cohosts the "Cultures of Energy" podcast (available on iTunes, PlayerFM, and Stitcher).

AKHIL GUPTA is Professor of Anthropology at the University of California, Los Angeles. His books Red Tape: Bureaucracy, Structural Violence and Poverty in India and Postcolonial Developments: Agriculture in the Making of Modern India attend to the power and arbitrariness of state development projects and practices in everyday life. He is currently working on a new ethnography about information technology workers in Bangalore.

PENNY HARVEY is Professor of Social Anthropology at the University of Manchester. Her main research interests are in the anthropology of science and technology, state formation and neoliberal regulation, and environmental controversies. She is the coauthor of *Roads: An Anthropology of Infrastructure and Expertise* and the editor of an edited volume on *Infrastructures and Social Complexity* (with Casper Brun Jensen and Asturo Morita). She is currently working on ideas of waste and the future in Peru.

BRIAN LARKIN is the Director of Graduate Studies and Professor of Anthropology at Barnard College, Columbia University. His research focuses on the ethnography and history of media in Nigeria. Most broadly he examines the introduction of media technologies into Nigeria—cinema, radio, digital media—and the religious, political, and cultural changes they bring about. He explores how media technologies comprise broader networked infrastructures that shape a whole range of actions from forms of political rule, to new urban spaces, to religious and cultural life. He has also published widely on issues of technology and breakdown, piracy and intellectual property, the global circulation of cultural forms, infrastructure and urban space, sound studies, and Nigerian film (Nollywood). He is currently completing the manuscript for *Secular Machines: Media and the Materiality of Islamic Revival*, which analyzes the role that media play in the rise of new Islamic movements in Nigeria and explores theoretical questions about technology and religion.

CHRISTINA SCHWENKEL is Associate Professor of Anthropology at the University of California, Riverside, and Director of the Program in Southeast Asian Studies. Her book *The American War in Contemporary Vietnam: Transnational Remembrance and Representation* examines historical knowledge production and the geopolitics of commemoration. Schwenkel has conducted extensive ethnographic research in Vietnam and eastern Germany. Her publications have focused on visual technologies of cold war memory and socialist reconstruction of urban infrastructure following the end of the U.S. air war in Vietnam. She is currently writing a book on architectural transfers and cultures of expertise, entitled *The Obsolescent City: East German Design and Its Afterlife in Urban Vietnam*.

ANTINA VON SCHNITZLER is Assistant Professor in the Graduate Program in International Affairs and an affiliate faculty member in the Department of Anthropology at the New School. Her research and teaching has focused on citizenship and political subjectivities, the anthropology of infrastructure and technology, liberalism and neoliberalism, colonialism and postcoloniality, energy politics, and South Africa. Her book *Democracy's Infrastructure: Neoliberalism, Techno-Politics and Citizenship after Apartheid* was published in 2016. Based on over eighteen months of archival research and fieldwork in Soweto and Johannesburg, the book is a historically informed ethnography of conceptions, technologies, and practices of citizenship and technopolitics at the contradictory juncture of liberation and liberalization in postapartheid South Africa.

68–77, 109, 121–25, 193–95, 224; sanitation and, 11, 28, 66, 133–35, 137, 141, 145; spatiality and, 14–15, 94–95, 139–40, 199n15; spectacle of, 106–8, 112, 134–35, 187; suspension and, 68–72; symbolism and, 26, 185–86, 195; teleology and, 9, 44–46, 54, 58, 62, 69–70, 81, 121, 141; telos and, 8–11, 70, 73, 81; time and, 12–20, 44–45, 47, 62–68, 82, 90–93, 98, 143; war and, 71, 74, 102–4, 119, 126n4. *See also* airports; bridges; capitalism; dams; highways; imperialism; liberalism; morality; temporality

Inka, 80
International Monetary Fund (IMF), 48–49
In the Time of Oil (Limbert), 54, 188
"Iron Construction F" (Benjamin), 9

Jacobs, Jane, 157
Jakobson, Roman, 176
Jimenes, Fernando Volio, 48
Jolobe, Zwelethu, 136
Joyce, Patrick, 4

Kaduna, 179–80
Kaplan, Robert, 200n22
Keck, Frédéric, 95
Kern, Stephen, 176
Keynes, John Maynard, 6, 205, 223–26, 235
Keystone XL pipeline, 229
Khan, Naveeda, 196
Kittler, Friedrich, 182
Knowing Capitalism (Thrift), 16
knowledge: classification of, 206, 208–9, 212, 217; forms of, 205–7, 212; humanities and, 207–9; institutions of, 215–16; interdisciplinary and, 207, 211, 215, 218; sciences and, 207–11, 216–17, 219. *See also* development; modernization
Knox, Hannah, 84, 86–87, 99n1, 199n17, 223
Kohn, Eduardo, 198
Koselleck, Reinhart, 81–82
Kracauer, Siegfried, 182
Kuhn, Thomas, 209, 218
Kumar, Krishan, 156
Kurke, Leslie, 204

labor and laborers, 43, 51, 86, 96, 102, 116, 126n14, 139, 227–29. *See also* farmers; workers
La Buena Esperanza, 43
Laclau, Ernesto, 148n1
Lagos, 195
Lakoff, Andrew, 95
Lansing, Stephen, 7
Larkin, Brian, 14–15, 19, 28–29, 44, 115, 134, 148n7, 167, 223–27
Latour, Bruno, 197, 209, 225
Lefebvre, Henri, 10, 197
LeMay, Curtis, 103
Lemke, Thomas, 21, 177
Lenin, Vladimir, 111, 231, 233, 237
Lerner, Daniel, 182
Levine, Caroline, 184
liberalism, 2–6, 22, 27–30, 141–42, 147, 148n4, 192, 235. *See also* development; infrastructure; neoliberalism
liberation movements, 140. *See also* race and racism
Life of Henry Brulard, The (Stendhal), 188
Limbert, Mandana, 54, 188–89
Lindsell, H. O., 179–81
Local Financial Accountability and Choice Act, 1

Mackenzie, Adrian, 177
Mail (newspaper), 134
Mains, Daniel, 119
Malabo, 41, 44, 47
Malema, Julius, 133, 146
Malm, Andreas, 233
Malthus, Thomas Robert, 211, 226
Mariam, Dossal, 158
Marres, Noortje, 169
Marvin, Simon, 10–11
Marx, Karl, 8–10, 24, 46, 65, 197, 205, 225–33, 236–37. *See also* socialism
materialism and materiality: aesthetics and, 177–79, 197; conditions of, 80–88; economy and, 197–98; engineering and, 92–98; form and, 23–28, 44, 104, 135–36, 175, 226–29; historical, 197, 212–13; hydraulics and, 156–70;

materialism and materiality (cont.)
  infrastructure and, 27–30, 49, 58, 83, 111–13, 145–46, 175–77, 197, 226; resources and, 6–8, 65–76, 97, 181; ruination and, 65, 120–25; symbolism and, 145–48. *See also* Bennett, Jane; Marx, Karl
Mattelart, Armand, 190
Mauss, Marcel, 7
Mba Abogo, César A., 41, 43–47
Mbembe, Achille, 48, 176
McFarlane, Colin, 10
Melley, Caroline, 199n21
Middle East, 224
Ministry of Culture and Education (Vietnam), 125
Ministry of Transport (Peru), 87
Mister Johnson (Cary), 190, 192
Mitchell, Timothy, 148n9, 203, 205, 224, 235
Mitra, Sumita, 77n1
Mode II science, 219
modernization, 7–9, 26, 46, 51–58, 64, 81, 122–23, 182, 192. *See also* development
Mol, Annemarie, 148n6
morality, 12, 22, 109–13, 120, 134, 161, 192, 238. *See also* development; infrastructure
Morgan, Lewis Henry, 9, 46
Morris, Rosalind, 150n28
Morton, Timothy, 239
Mràzek, Rudolf, 179
Mumbai, 156–58, 164, 167–68
Mumbai Municipal Corporation, 160–61
Municipal Corporation of Greater Mumbai, 155

Narita Airport, 71, 74–75
National Science Foundation, 211
Native Nostalgia (Dlamini), 149n20
Nehru, Jawaharlal, 64
neoliberalism, 2–6, 123, 170n1, 223–26, 235. *See also* liberalism
New Deal, 6
Ngai, Sianne, 175
Ngwenya, Eunice, 144
Nielsen, Michael, 216

Nigeria, 182, 190–92
Northern Provinces of Nigeria (SNP), 179, 198n3
Nye, David, 1–6

Obiang, Nguema Mbasogo, 48, 53, 57–58, 59n1
Oil Control Center (OCC), 230
Olympic Games, 64
Oman, 188–89
OPEC, 224
Operation Gcin'amanzi (Operation Save Water), 143, 145
Operation Vul'amanzi (Operation Let the Water Flow), 143
Opium Wars, 159

Packer, George, 200n22
Paris Climate Accord, 29
Pasteur's Quadrant, 219
Peru, 3, 28, 80–87, 90–95, 99n1. *See also* Cusco
Phiri Concerned Residents Forum, 143
pipelines, 20, 29, 166, 212, 230–32, 236. *See also* fossil fuels; infrastructure
Plato, 184, 213
politics: aesthetics and, 176–77, 183–84, 186–87, 190; of apartheid, 28, 148, 148n13; economy and, 86; form and, 83–84, 175, 184–86, 196–98; governance and, 180–81; local, 161–68; technology and, 14; technopolitics and, 4, 13–14, 109, 133–39, 141–47, 148n9. *See also* biopolitics; capitalism; development; modernization; neoliberalism
poo wars, 133–34, 146, 148n5. *See also* South Africa
postcolonialism, 28, 64, 102–7, 135, 157
posthumanism, 225
postmodernism, 5, 15
poverty, 141, 156, 162
Povinelli, Elizabeth, 148n6
privatization: apartheid and, 134, 140–43; hydraulics and, 22–23, 155–57; infrastructure and, 28, 42, 48–52, 57, 66, 134, 155–57, 161, 176–77; publics and, 5,